PHYSICS RESEARCH AND TECHNOLOGY

QUINTESSENCE

A THERMODYNAMIC APPROACH TO THE PHENOMENA OF NATURE

PHYSICS RESEARCH AND TECHNOLOGY

PHYSICS RESEARCH AND TECHNOLOGY

QUINTESSENCE

A THERMODYNAMIC APPROACH TO THE PHENOMENA OF NATURE

LEV Z. VILENCHIK, PHD

nova science publishers
New York

Library of Congress Cataloging-in-Publication Data

Names: Vilenchik, L. Z. (Lev Zalmanovich), author.
Title: Quintessence : a thermodynamic approach to the phenomena of nature / Lev Z. Vilenchik, Ph. D., Senior Consultant, Felicitex Pharmaceutical, Inc., Cambridge, MA, USA.
Other titles: Physics research and technology.
Description: Hauppauge, New York : Nova Science Publisher's, Inc., [2016] | _2016 | Series: Physics research and technology | Includes bibliographical references and index.
Identifiers: LCCN 2015050184 (print) | LCCN 2016003078 (ebook) | ISBN 9781634844475 (hardcover) | ISBN 1536122435 (softcover) | ISBN 9781634845250 (ebook)
Subjects: LCSH: Philosophy of nature. | Thermodynamics--Philosophy. | Life (Biology) | Cosmology. | Entropy.
Classification: LCC BD581 .V55 2016 (print) | LCC BD581 (ebook) | DDC 501--dc23
LC record available at http://lccn.loc.gov/2015050184

Published by Nova Science Publishers, Inc. † New York

CONTENTS

PREFACE

The manuscript describes a fundamentally novel approach to understanding the phenomena of Nature from the standpoint of classical, relativistic, quantum physics, and cosmology.

Regularities of the living and inanimate worlds, as well as the interaction between the two and within each of them, are described there. It is shown that for every phenomenon (even seemingly supernatural) one can find a scientific interpretation. In the words of Albert Einstein: "The scientific search is a constant race away from the magic."

What kind of regularities are we talking about?

First of all, they are regularities of the phenomena well-known to us from different branches of physics. Among them there are: gravitation, phase transitions, heat conductivity, diffusion, quantum entanglement, and many others.

These regularities are considered in the book from a new point of view which does not cross out the laws of Nature known to us, but explains their existence, proceeding from their deeper understanding. Four fundamental interactions (gravitational, electromagnetic, strong and weak) are treated in the book as affiliated from entropic interaction, whose concept, in principle, is not new, but it is usually used in a subjunctive mood. ("macromolecule links, as if, entropically repel from each other at a short distance and entropically are attracted to each other at a long distance"). Instead of this, the entropic interactions are considered as a quintessence of Nature. A description is provided of how the entropic order rules the World, having organized in a peculiar way the distribution of matter in the Universe, its intrinsic laws, the eternal DNA in our perishable bodies, and the moral codex that we follow and the introduction of real-life entropic interaction offers further explanation of the regularities, establishing a correlation between the change of entropy and a rhythm of time, allowing an extension of the laws of physics to public life, art, literature, religion and human beings. Such an approach is also a peculiar bridge between physics, medicine and its alternative options. It makes recommendations for the treatment of diseases and the prolongation of our lives.

Being applied to human history, the approach tries to explain the mysterious disappearance of the Mayan civilization in Central America (highly advanced for its time) and the possibly related puzzle of Easter Island.The book discuss how the conditions needed for the emergence and development of organic life on Earth were created. Why was the chosen set of approximately 100 global constants, which govern our life, presented to us by Nature? Are there other planets in the Universe where there is also intelligent life? And how could contact be established with it, if there is any?

The book offers recommendations on how to achieve desired outcomes in daily life, including solutions to health issues, longevity, and simply improving one's quality of life.

The book is written in the popular science format to be comprehensible to the general public. It does not require its readers to have a special math or physics background. However, for those mastering in mathematics and physics, a detailed mathematical description is included in the Supplementary sections.

The proposed approach claims a lot but in fact is quite simple. No words could better reflect its essence than a quote by Antoine de Saint-Exupéry: "Perfection is achieved not when there is nothing to add, but when there is nothing to take away."

Dr. Lev Vilenchik
Senior Consultant, Felicitex Pharmaceutical, Inc.
Cambridge, MA, USA
82 Bishops Forest Drive, Waltham MA 02452
Email: lev.vilenchik@gmail.com

INTRODUCTION

QUINTESSENCE

"Two things awe me most,
the starry sky above me and the moral law within me."

Immanuel Kant

The Entropic order rules the World,
Having organized in a peculiar way
The Distribution of matter in the Universe,
Its intrinsic Laws,
The eternal DNA in our perishable body,
And the moral codex that we follow

Quintessence (from Latin: quīnta essentia - fifth essence) - a multiple-meaning term.

Quintessence in ancient and medieval natural philosophy and alchemy is the fifth finest element, additional to the four main elements, which were considered to be earth, water, air and fire.

In modern physics the term “Quintessence” is used sometimes as a hypothetical form of dark energy, postulated to explain the accelerating rate of the Universe widening.

In religious literature it means: “The Soul that permeates the whole world, the spirit of the world, a great creative force, which God has given to the material world.”

In the modern conventional sense, Quintessence means the main sense, the most important, the most significant, the main essence. It is this interpretation of the term “Quintessence” that is used in this book, though the reader, at his/her discretion, may in some cases use other meanings of that multiple-meaning term.

Chapter 1

EXPLANATION OF PHYSICAL REGULARITIES FROM POSITIONS OF THE THERMODYNAMIC APPROACH

FROM THE AUTHOR

"Perfection is achieved not when there is nothing to add,
but when there is nothing to take away."
Antoine de Saint-Exupery

Nature amazes us with its versatility. It is full of the most amazing phenomena, the essence of which is outside current scientific knowledge. We try to understand these phenomena, to discover the laws underlying them, and to use this knowledge in our daily lives. We already understand that Nature appears to us at three different levels. At one level, Nature shows us all that we can see with naked eyes and feel using our other four senses. At another level, it opens the microcosm into which we penetrate with the help of devices created by us. And at the third level, it strikes our imagination with regularities manifested at large distances, high speeds or long time intervals. In addition to these levels, it surprises us - sometimes with stunning similarity, and sometimes with fundamental differences, of the two closely intertwined worlds of living (organic) beings and inanimate (inorganic) objects.

Usually, we are captives of our current understanding of the nature of things. It may be difficult to change it, like an old-fashioned garment to which we have become too accustomed. Even a genius such as Albert Einstein, who in his theory of the photoelectric effect [1] gave us the first impetus for the development of quantum physics, to the end of his life could not abandon the determinism of classical physics. He said he did not believe "that God plays dice," while keeping in mind the probabilistic interpretation of the laws of Nature as proposed by quantum mechanics. But it was Einstein who destroyed the classical physics notions established by many centuries.

Relativistic Theory (Special [2], and then the Total [3]) replaced the well-familiar 3-dimensional world with the 4-dimensional one. Distance and time in each physical system, according to the theory, began to depend on the relative velocity, and the flat space (with Euclidean geometry) turned into a bent space, where the non-Euclidean geometry exists and parallel lines intersect. Gravity was interpreted as a consequence of the curvature of the space. "I am like a beetle, who realized that it was crawling along a curved branch," - Einstein explained the essence of his discovery to his son.

Figure 1. Niels Bohr and Albert Einstein at a scientific seminar.

Figure 2. A. Einstein speaking at scientific conference.

Figure 3. An analogy of the non-Euclidean space in which we live, proposed by Einstein.

The scientific world accepted Einstein's discovery and, after nearly a century, accepted the explanation for the birth of our Universe given by Alan Guth, who assumed negative gravity [4] - i.e., instead of attraction, there was the repulsion of gravitational masses at the moment of the Big Bang [5]. Nobody ever observed negative gravity, but people believed in it, as well as in the origin of material objects from a vacuum. Being a theoretical physicist who originally specialized in the quantum field theory (publication in Vestnik of Leningrad State University, 1967, "Construction of Quantum Field Theory with the Measures in Functional Spaces"), the author was able to confine this book within the well-known provisions of classical physics (mainly thermodynamics) to explain different phenomena of Nature, including those which now are attributed to the category of mysteries. The author tried to avoid mathematical calculations, though he believes that physics without mathematics looks like an army without weapons. So, in order not to leave this army completely unarmed, unable to convincingly prove its strength and arguments, the author had to include a few mathematical formulas in the text. However, the reader, if not educated in mathematics, may omit them, going directly to the conclusions following them. More detailed mathematical calculations are rendered in the appendixes.

The thermodynamic approach to the phenomena of Nature proposed in the book is based on experimental facts. It looks very simple, but it claims much. The relationship between the rate of entropy change in a thermodynamic system and the rhythm of time is investigated in this approach. In particular, a possible explanation for the short-term emergence of negative gravitation at the time of the Big Bang is offered here. Interactions between objects in the world of animate and inanimate Nature, including the psychological effects connected with the consciousness of the human being, are considered. Much attention is paid to the medical aspects of a condition of a living organism. It appears that corresponding sections will interest physicians and pharmacologists and will give a push to the development of new methods for the treatments of various diseases. If in Sigmund Freud's psychoanalysis [6] much in-person behavior is rooted in a person's early childhood, so in the thermodynamic approach the determining factor is a kind of entropic order that exists in the human body and in the surrounding Nature. A statement in an epigraph from Antoine de Saint Exupery: "Perfection is achieved not when there is nothing to add, but when there is nothing to take away" could not better reflect its essence. Hopefully, this approach will become interesting to the scientific community and to the wider readers of this book.

I. Preamble
The Principles Used in the Description of the Laws of Nature

Having barely learned to speak, the child showers the adults with questions:

"Why, why, why? ..."
Why is it light in the daytime and dark at night?
Why is the sky blue, and the leaves are green in the trees?
Why are summers hot and winters cold?
Why does a balloon fly up when released?
And so on and so forth.

As the child get older, the number of issues first grows like an avalanche, and then falls off and is limited to the specific characteristics of the acquired knowledge. And only a few people, before their hair turns grey, continue to be surprised by the amazing diversity of Nature around us and try, now themselves, to find answers to the many questions tormenting them (including some of the questions from childhood) which still remain unanswered. This is true - it is God's gift to keep a child's curiosity and an ineradicable desire to understand the world around us, having penetrated into its very core, into its quintessence.

Indeed, why does a balloon tend to fly? Well, first of all, not every balloon does this, but only those filled with light gas or hot air. And, secondly, the balloon tends not to fly anywhere, not in any direction, but upwards. This means that the net force acting on it is directed, specifically, upwards. What is this power?

The air molecules surrounding the balloon "are knocking" from all directions. And the fact that the ball flies up means that a bigger number of the molecules strike it from below, rather than from above, and that the impulses of the "lower" molecules are, possibly, larger than the impulses of the "upper" molecules. In reality, air molecules are distributed in the gravitational field of Earth unequally. Their density and impulses decrease with the increase of the distance from Earth (according to the Boltzmann and Maxwell distributions, respectively, [7]).

As a result, the air tends to push the balloon up and gravity pulls it down. By vectorially developing, these two forces are directing the balloon upward if the pushing-out force surpasses the balloon's weight. Otherwise, it drops down. If these two forces are equal, the balloon remains hanging in the air. So, if you imagine that the shell of the balloon is weightless and the balloon is filled up by the air that is around the balloon, then (in the absence of any directional movement of air) the balloon will remain stationary and it is easy to conclude that the buoyancy force is exactly equal to the weight of the air displaced by the balloon.

Figure 4. A balloon filled with light gas flies upwards.

Figure 5. Archimedes, rejoicing at making a discovery.

Similarly, this explains the buoyancy of bodies in any liquids. Only in this case, the air is replaced by the liquid, and its molecules transfer their impulses, fluctuating about a position of equilibrium and creating different hydrostatic pressure on the bottom and the top surface of a body immersed in fluid.

Figure 6. Discovery of Newton's law of gravitation.

This phenomenon is known in science as Archimedes' principle. Legend tells us that Archimedes discovered his law while taking a bath. He felt that his body in water weighed less. With a cry of "Eureka" (which means: I have found), he jumped out of the bathtub and ran naked through the city, unable to contain his enthusiasm for his discovery.

The same can be said about another legend, according to which Newton discovered the law of gravity after an apple fell from the tree under which he was sitting, and hit his head. Like everyone else, he could see different objects falling, or their sliding down a sloped surface, and could logically understand the phenomenon, finding its cause.

Both Archimedes and Newton, in their reasoning, could start from the same principle - the principle of causality [8-10]. This is an axiomatic principle that is taken for granted without proof, because it seems obvious, but it is not clear how it can be theoretically proved. Its essence is that every phenomenon is preceded by its cause. All phenomena are consequences of certain reasons. And of course, they are based not only on the principle of causality, but on a long series of other axiomatic principles, the truth of which has to be believed by us because of the apparent impossibility of their proof. This axiomatic approach is widespread in the natural sciences. For example, Euclidean geometry (according to D. Gilbert [11]) is built on 20 axioms, the most famous of which is the axiom about parallel lines, stating that they never intersect.

Among the axiomatic principles that will be used in this book, in addition to the principle of causality, there are included: Mach's Principle [12], Le Chatelier's Principle [7], the principle of least action [13] and three principles (laws) of thermodynamics.

The principle of causality, in addition to what has been said earlier, states that for two events divided in time, the earlier one can have an impact on the latter event, but not vice versa.Mach's Principle in its most general form [5] postulates that local physical laws are determined by the large-scale structure of the Universe. This means that the processes occurring in any local region of space depend strongly on the processes in a wider area, as well as on the characteristics of the entire Universe, processes occurring in it, and related changes. The converse is true, i.e., processes occurring in the local systems provide a possible impact on the systems in which they participate. According to E. Mach [12], any changes in any part of the Universe change the state of the entire Universe, and have a corresponding impact on all of its other parts. Albert Einstein emphasized [14] that Mach's Principle had played an important role in the creation of the general theory of relativity, in which all existing material bodies (from small objects to galaxies) are constantly interacting by means of their gravitational fields, which fill the entire Universe.

Le Chatelier's Principle declares that all systems in Nature are inertial. In other words, they oppose any changes to which they are forced by external influences, i.e., at an external influence that upsets the system out of some steady state, processes seeking to weaken the results of this influence appear in the system.

The principle of least action in physics occupies a special position. All of the currently known fundamental interactions - electromagnetic, strong, weak, and gravitational ones - obey that principle. The basic equations of classical and quantum mechanics, electrodynamics, quantum field theory, thermodynamics and of the general theory of relativity (gravity) follow from it [15-25]. According to that principle, all changes in Nature must be such that its certain quantitative characteristic (usually associated with energy) remains (or at least tries to stay) minimal. The principle of least action tells us that in Nature there is nothing superfluous; processes occurring in it do not overlap,

Nature is economical, and for any of its evolution, minimal expenditure of energy is enough.

All three principles (or laws) of thermodynamics are axiomatic. They are not theoretically proven and are based only on experimental facts [16].

The First Principle of Thermodynamics is the law of energy conservation. It says that energy is not created, does not disappear, and only passes from one sort into another.

The Second Principle of Thermodynamics claims that the function of a state of the isolated thermodynamic system (called entropy) always increases.

The Third Principle of Thermodynamics says that as the temperature approaches absolute zero (-273^{0}C) the entropy of the equilibrium system does not change.

Those principles, and the laws of Nature deriving from them, will apply to all of the phenomena of the organic (living) and inorganic (inanimate) world considered in the book. This approach is based on the belief that the laws of Nature are common to both worlds; the organic world is only more organized, processes in it are speedier and therefore the organic world is more volatile and evolves much faster than the inorganic world.

II. Theoretical Prerequisites of Application of the Thermodynamic Approach to Natural Phenomena

1. Stochasticity

"Necessity makes its way through the crowd of accidents."
Philosophical truth

The subject of research in physics is physical systems. They consist of material objects. When a physical system (which has a large number of degrees of freedom) is considered as a whole without the involvement of the characteristics of its constituent objects, it is called a thermodynamic system. These systems include solid, liquid and gaseous objects, individual macromolecules, galaxies, and the Universe. Any object of living nature, including humans, also fits this definition.

According to Mach's Principle, as described in the preamble, all objects of

Nature are interconnected. They exert a certain influence on each other, regardless of the distance separating them, i.e., they interact. Remote objects interact more weakly; neighboring objects interact strongly. The main scientific discipline describing the interaction between material objects is physics. In classical mechanics, objects interact through direct contact. In electrodynamics and the general theory of relativity, objects affect each other by means of fields (electric, magnetic, electromagnetic, gravitational). In nuclear physics, strong and weak interactions work. In thermodynamics, statistical physics and quantum mechanics, in comparison with classical physics, there are additional characteristics of the interaction: thermodynamic potentials, entropy, enthalpy, and the wave function, and the physical laws gain a probabilistic (stochastic) nature.

We will consider mainly the thermodynamic systems, so it makes sense to define their basic characteristics.

Enthalpy, H, is a measure of the total internal energy of a thermodynamic system. In particular, the change in the H at the processes occurring at constant pressure equals the amount of heat produced by the system.

In classical thermodynamics, entropy, s, is defined as a definite physical quantity. It can be measured just like the length of a rod, the temperature of an object or amount of heat in it. Entropy is a function of the system state, which is equal to the amount of heat that the system receives during an equilibrium process, or gives off, referred to the system temperature in the absolute scale: $\delta s = \delta Q/T$, where δs is entropy change; δQ is the change of the quantity of heat system, and T is the absolute temperature of the process.

Entropy establishes a link between macro- and micro- states. This is the only function in physics that shows the direction of processes. Since the entropy is a state function, it does not depend on how the transition of a system happens from one state to another one. The entropy value depends only on the initial and final states of the system.

From the other side, (which is especially important for us) entropy, s, characterizes the degree of order in the system, the degree of homogeneity. The more the system is ordered (structured) and is non-uniform, the less is its entropy. From the point of view of statistical physics, entropy is the logarithm of probability of the state of a system taken with the negative sign (Boltzmann's H-theorem). It is a positive countably additive function, i.e., it is the sum of entropies of separate parts of a system (subsystems). If the closed system isn't in a condition of statistical equilibrium, it eventually passes into more and more equilibrium states, i.e., more probable states, being characterized by smaller orderliness, bigger uniformity and, therefore, a higher value of entropy. This law of entropy increase is the Second Law of thermodynamics. An essential feature of entropy is the following: if the temperature of an equilibrium system approaches absolute zero, its entropy does not change (the Third Law of thermodynamics).

Free energy, G = H - Ts, is the energy of a system which can be used for producing work. T is the absolute temperature according to the Kelvin scale.

The state (or condition) of a thermodynamic system is understood as a set of all its characteristics. A change in at least one of them means a change in the condition of the whole system.

As has already been said, the evolution of thermodynamic systems and processes (physical, chemical, and biochemical) occurring there have a significantly stochastic character. The main characteristic of those systems is the value of their free energy, G. At an equilibrium state the G value is a minimal one. If the system is at a nonequilibrium state, it seeks to reach an equilibrium state, lowering the free energy to the minimal value inherent in it. And, as the free energy of the system consists of two components, enthalpy and entropy, at the description of processes occurring in the system we have to take into account the corresponding entropy and enthalpy changes occurring in that system. It is impossible to say whether the thermodynamic system is in a certain or in another state. Instead, it is necessary to speak about the probability of its staying in that state (with an index of 1) and about the probability of its transition to another state (with an index of 2). The probability of that transition $W_{1,2}$ is unambiguously defined by occurring enthalpy and entropy changes,

$$W_{1,2}=exp(-\Delta G/kT), \quad (1)$$

where ΔG is the change of the free energy G of the system, and T is its temperature.

$$\Delta G=\Delta H-T\Delta s \quad (2)$$

2. The General Tendency in Evolution of Thermodynamic Systems

How does the evolution of an isolated (closed) thermodynamic system proceed? According to the First Law of thermodynamics, its energy cannot change. But certain processes are occurring in the system. They are responsible for the evolution of the system. The processes alter the characteristics of the system. The system evolves. What is the driving force of those processes? The answer is clear: it is the tendency of the system to move to a more probable state, i.e., to reduce its free energy, which, as we know, consists of two components: an enthalpic part and an entropic one. But if the enthalpic part, representing the internal energy of the system, must remain constant, there is only one option - the change of the entropic part which is the product of entropy of the system and its temperature in the absolute scale, i.e., to reduce the free energy of the system, its entropy must increase, as required by the Second Law of thermodynamics. In other words, the system seeks to increase its entropy. And how can it be done?

The answer is the following: By "turning on" physical, chemical, or other processes which will change the state of the system, and will increase the value of the entropy. And at the heart of all processes there are interactions between the objects that make up the system. They can be both macro- and micro- objects. Thus, because of the tendency to increase its entropy, the system "turns on" these interactions and the processes generated by them.

The simplest examples of that tendency are well-known processes of heat conductivity and diffusion. In both cases a thermodynamic system tends to decrease its free energy by increasing its entropy. The heat spreads in the system by means of the process of heat conductivity. And substances spread by means of diffusion.

It should be noted that the minimum of free energy of Gibbs, G, corresponds to the condition of stable equilibrium of a thermodynamic system. Any external influence causes its perturbation and related structuring. Thus, the conditions of the system, its enthalpy, entropy, and free energy, are changing. The inertia of Nature, according to Le Chatelier's Principle, stimulates in the system some processes that seek to weaken the results of the external influence. For example, if as a result of this influence the entropy component of the free energy of the system, sT, is decreased, and the value of free energy, G, is respectively increased, then in the system - or in any part of it - arises processes seeking to reduce the free energy and to return its initial value. Those processes lead either to an increase in the value sT, or to an expense of the internal energy of the system, i.e., to the fall of its enthalpy, H, as reflected in formula (2). At that, the increase in the value sT can occur due to the growth of either its entropy, s, or its temperature T, i.e., the thermodynamic system seeks to compensate the change of its entropy component of the free energy by the corresponding change of its enthalpy and vice versa.

That tendency is well-known in thermodynamics and without looking into the details of it can be easily received from the principle of least action which, as we already noted, is one of the key provisions of modern physics. In Appendix 1 we give detailed consideration (from that principle's positions) of the behavior of the thermodynamic system which is in the condition of rest in the chosen reference system.

3. Entropic Interactions

A. Heat Conductivity and Diffusion

Heat conductivity and diffusion have already been mentioned as instruments of the evolution of a thermodynamic system. They can also be considered as examples of the entropic interaction between different parts of the system. Indeed, the entropy is different in parts with different temperature and composition. So, we observe a tendency for the system to be homogeneous according to the distribution of heat, substances and entropy. That is, we can conclude that this tendency is a result of the entropic interaction between parts of the system.

B. Stabilization of Room Temperature

Let us consider another example. Imagine in the room there is a vessel with liquid, which, together with the air surrounding it, can be considered a thermodynamic system. And imagine that for some reason the air temperature in the room goes down to, or below, the freezing temperature of this liquid. (As a possible reason for this, we can specify a fall in the temperature of the walls at an invariable, or nearly invariable, amount of heat in the room.) What will occur? First of all, with the fall of temperature the entropy of the air will increase. This means that the entropy of the system, liquid-air, will increase at the same time and the system will be removed from an equilibrium state. According to Le Chatelier's Principle, processes (or at least one process) lowering entropy have to start in the system. In this case, it will be the freezing of the liquid. The entropy of the liquid decreases at the formation of ice.

However, the balance of the system - entropy and internal energy (i.e., enthalpy) - is thus broken. For restoration of this balance, the liquid during the freezing gives a part of its internal energy, i.e., heat, to the air surrounding it. The temperature of the air will rise, the entropy will go down and the whole system, liquid-air, comes back to its initial stable state.

An opposite process works if the air temperature in the room increases above the melting point of the liquid. The air's entropy goes down. Ice (from a previously frozen liquid) thaws, increasing its entropy, taking away excess heat from the air and lowering the temperature of the air. As a result, the initial stable state is restored. The freezing of liquid and the thawing of ice are considered in classical thermodynamics to be a result of the heat exchange of a liquid with the environment. Entropy changes occurring are thus considered secondary and are treated as a consequence of that heat exchange. Actually, the entropy changes precede the above-mentioned heat exchange. As soon as the entropy of the air surrounding the liquid changes, the entropy of all the system (liquid-air) changes at the same time. This means that the information on the change of the entropy of air becomes instantly known to this liquid, thanks to the entropy signal system, which should be put into practice for all similar cases. And already, as the result of the information received thus and the related entropy interaction between liquid and air, a heat exchange begins in the system (between its various parts).

C. Energy Preservation

This heat exchange is used for cooling and heating rooms without energy expenses. Solutions of hydrates of salts are the cheapest and most available for these purposes. Their temperature of freezing/melting significantly depends on the concentration of the solution. With concentration growth, this temperature grows, and the range of its change lies between 0°C and 40°C. At the same time, the corresponding change of an enthalpy, i.e., heat radiation or absorption, for these solutions is very high. For example, for sodium phosphate containing 12 molecules of water ($Na_2HPO_4 12H_2O$), the melting enthalpy is equal to 280 kJ/kg at a temperature of 35°C. The heat which is *emitting* at the freezing of 1 kg of this solution is enough to heat or cool air by 1 degree Celsius indoors of 100 square meters and 3 meters high.

By selecting the concentration of the solution, it is possible to lower the temperature of melting or freezing to a comfortable one (for example, 23°C). Then the temperature fall at a cold time of the day (or a season) can be compensated by the presence in the room of several "batteries," each containing, say, 10 kg of sodium phosphate solution. Respectively, at a warm time, when the temperature increases to above 23°C, the solution will start thawing and taking away heat, i.e., to cool the room.

This phenomenon is widely used in tropical countries (for example, in India) for the cooling and heating of houses. It would be good to use it to maintain a constant temperature in the rooms of nuclear power plants; that would reduce the possibility of accidents associated with the increased temperature. Thus, there would be no need to use any temperature sensors, which under certain conditions can fail.

D. Weather Forecasts

Coming back to formula (1), it is necessary to emphasize that the stochastic nature (observed by us) of the change of conditions of thermodynamic systems is not connected with the incompleteness of our knowledge of them. Its essence is in the probabilistic nature of the physical laws which are being realized here. It is visually illustrated, for example, in the weather forecasts that we receive daily. When we are promised a 70% probability of rain, it does not mean that meteorologists do not have enough information about the condition of the atmosphere. The latter is a thermodynamic system and the change of its state, in principle, is described by formula (1), although in practice it is calculated on the basis of other reasons.

The same can be said about the behavior of any person and about events accompanying his or her course of life. For example, despite having an accurate road map and directions, a person can be late to a planned meeting place because of road incidents on the way, an unexpected change to the train schedule or any other accidents. However, with a certain persistence, the person's intention to appear eventually in the planned meeting place can be carried out. Do you remember the phrase from the epigraph, that necessity lays itself a way through a crowd of accidents? And so, it occurs pretty often in everyday life.

The behavior of such thermodynamic systems as the ideal gas or the diluted solutions of low-molecular substances and the flexible-chain macromolecules is a fine illustration of entropy interactions at a molecular level. The random processes occurring in these systems realize entropy interactions at the molecular level and also send systems to more probable (and more equilibrium) states.

4. Flexible-Chain Macromolecules

Macromolecules consist of a large number of molecules which have a small molecular mass. These small molecules may be considered repeating links of one macromolecular chain. The number of the links may come up to tens and hundreds of thousands and even up to millions of units.

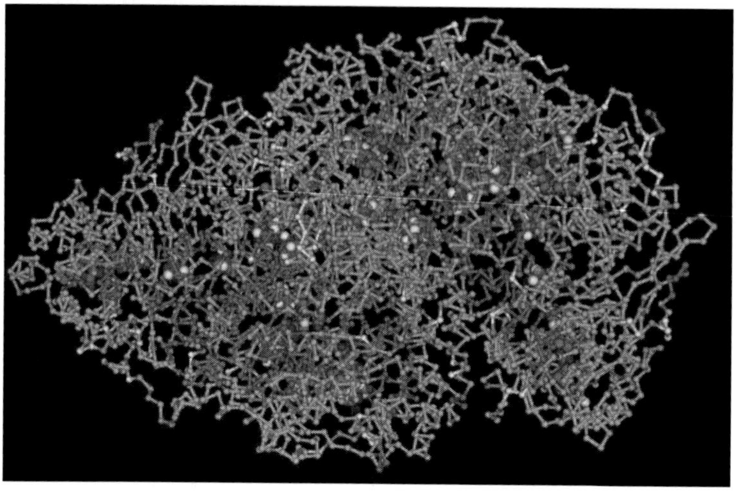

Figure 7. Illustration of a macromolecule of polypeptide.

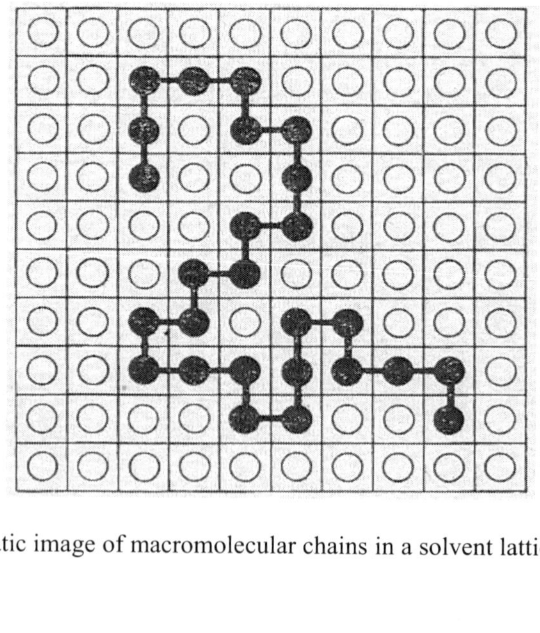

Figure 8. Schematic image of macromolecular chains in a solvent lattice [28].

In the case of flexible-chain macromolecules in ideal dilute solutions, the entropic interaction of the macromolecular links appears as an entropic repulsion for neighboring links and it appears as an attraction for links which are far from each other. This interaction is realized by means of a random walk of the macromolecular links. This fact is used in so-called lattice models at calculations of the entropy of macromolecules [26, 27]. The solvent is simulated (in the models) as a regular lattice. Each cell of the lattice includes one molecule of the solvent or one link of the macromolecular chain (as shown in the Figure 8). A random walk of the macromolecular links, realizing their entropy interaction, occurs under the influence of impulses received by the macromolecular links from collision with the solvent molecules. The solvent molecules are in a continuous random walk as well because their electromagnetic radiation is in the infrared range. Photons from this radiation are going randomly in all directions. Jet return forces molecules to move in the opposite direction. Besides the radiation, the solvent molecules capture photons radiated by neighboring molecules. Absorbing these photons, molecules receive the corresponding impulse summarizing with an impulse which they have at the time of the absorption.

The solvent molecules are distributed inhomogeneously in the macromolecular solution. The density of the molecules is higher in the area between macromolecular links which are close to each other (because of solvent molecules solvated on the macromolecular links). Therefore, the summarized impulse received by the neighboring links is directed out of the area. The situation is the opposite for links which are far from each other. As a result, links are distributed in space in such a way that their density is maximal in the geometrical center of a macromolecule and exponentially decreases in the process of moving away from it. So, a certain average conformation, an average size and an average form which are considerably defined by the entropy effects [29], may be collated for a flexible-chain macromolecule. Thus, the links of a macromolecule do not transfer to each other any information on their distribution in the solution. A change in the position of any link means a simultaneous (instant) change of the conformation (and consequently of the entropy and also of Gibb's energy) of the whole macromolecule.

So, the change of the thermodynamic state of any part of the system is the state change for the whole system, but processes caused by this change (in the given case they are: electromagnetic radiation, random walk of the solvent molecules, transfer of the impulses to macromolecular units) require some time interval for the starting and realization of the processes.

Many regularities which appear in ideal gases and the diluted solutions of low-molecular weight substances can also be explained as a result of an entropic interaction of molecules. This is manifested as their mutual pushing away or attraction, which is carried out at the expense of forces of Van der Waals [30].

It should be noted that, as in the case with macromolecules, the chaotic movement of these molecules is influenced by their distribution in space. If this distribution is non-uniform, as a result of radiation and absorption of photons, the molecules, continuing a random walk, will preferably move in the direction of areas where their density is less and where there are more empty places in space. This results in their more uniform distribution in space and the entropy growth for all their sets, i.e., one might say that molecules of gas or solution (as well as of any other material objects) entropically repel from areas with a high density of their distribution, seeking to get to areas with a smaller density.

If we take into consideration the gravitational (or any other) interaction between molecules, which brings their mutual attraction, it will essentially change nothing. The energy of an attraction is negative. It reduces the enthalpy value and consequently also the Gibbs's free energy, transferring a set of considered molecules to a more probable state. If the force of this attraction is more than the force of the entropic pushing away, material particles (in the absence of other forces) approach each other. Otherwise, they repel and move away from each other.

It is known from nuclear physics that the neutrons in an atomic nucleus repel each other at short distances (significantly less than 10^{-13}sm), while at greater distances (about 10^{-13}sm or more) a specific strong interaction leads to their mutual attraction. The general tendency to an increase of entropy dictates a tendency to the uniform distribution of molecules in a thermodynamic system. From here follows (for limited volumes) the pushing away of molecules (and of any other material objects) from each other at short distances and their attraction at greater ones. If objects are distributed in an unlimited space, their entropic pushing away prevails over a gravitational attraction which leads to the continuous expansion of any gas cloud and of all of our Universe as a whole.

5. Quantum Entanglement

Entropic interaction and its related entropic way to transfer information can be used for an additional explanation of the quantum entanglement phenomenon [31]. This phenomenon is observed when quantum states of the spatially divided particles (which are entangled particles) instantly change (in a strictly definite way) when the state of one of them changes. The spatial coordinates of particles, their impulses, the moments, polarization, etc. can be the measured characteristics of a state. Thus, when the characteristic of one of the entangled particles is measured in an experiment (for example, own mechanical moment (spin), focused in the clockwise direction), and then the spin of another entangled particle is measured, the latter one is always focused counterclockwise. This correlation is observed as occurring instantly after the first measurement, i.e., at the same time and irrespective of the distance dividing these entangled particles. Such behavior naturally follows from a formalism of quantum mechanics, but it is impossible from the point of view of classical physics, which caused in 1935 a sharp discussion between Einstein and Bohr [32].

The numerous experiments made since then [33-35] have confirmed with high reliability the conclusions of quantum mechanics. The last confirmation resulted from recent experiments carried out in *the Netherlds* by the international group of scientists led by Dr. B. Hensen. The results of the experiments ("Loophole-free Bell inequality violation using electron spins separated by 1.3 kilometers") were published in "Nature" online journal on October 21, 2015. Moreover, earlier results showed that information on a change of state of one of thef entangled objects becomes instantly known to another object (or group of other objects), not only in a case where these objects are elementary particles (photons, electrons, etc.), but also with molecules having the size of fullerenes (buckyballs) [36] and even of small diamonds.

From the standpoint of the thermodynamic approach outlined in this book, the entangled particles make up one thermodynamic system. Each of these particles is a part of this system. When you change the state of a particle, its entropy changes as well. Simultaneously, it is also a change of state (entropy) of the entire system, i.e., all other components of the system of the entangled particles or objects react instantly to this change accordingly.

Figure 9. Albert Einstein and Niels Bohr discussing the problem of a quantum entanglement.

In principle, all objects of Nature that are open thermodynamic systems can be regarded as entangled objects. An irreversible change in the state of any of them is the change of the entropy of the object and is at the same time a change in the entropy and thus of the state of the totality of the objects. That is, we can say that any irreversible changes in open thermodynamic systems simultaneously lead, according to Mach's principle, to corresponding changes in the entire Universe. Such an interpretation of Mach's principle seems preferable to me. The usual lever scales can serve in classical physics as a model analogy of the entangled objects. The lowering of one of their bowls under the influence of the weight "instantly" leads to the lifting of another bowl.

6. Chromatography of Macromolecules

Addressing once again the example of macromolecules, one should note not only the mutual pushing away or attraction of their links, but also the entropic pushing away in which a macromolecule participates as a whole object. If we place a membrane that is impenetrable for macromolecular links into a solution, the probability of finding macromolecules near the membrane turns out to be less than at a considerable distance from it. In the space between two such membranes placed in parallel to each other at some distance, the probability of the presence there of macromolecules will be much less than in a free solution (without membranes where the probability is equal to 1 [36].

In Appendix 2 it is shown that the narrower the space between membranes is, the lower the probability of a macromolecule being there. The probability of the detection of links of a macromolecule near walls is always less than at some distance from them. Thus, it is possible to say that the macromolecule as a whole has an entropic repulsion from impenetrable walls.

Many of us feel a sense of discomfort at being in a small closed room. For some people, this discomfort reaches even pathological fear, considered in medicine as the disease claustrophobia (because of which such people cannot serve, for example, on submarines). In the same category of sensations also belongs the instinctive desire, while driving, to stay away from cars moving in the adjacent lane on the road, or from a fence alongside the road. This should be classified as a manifestation of the entropy repulsion.

One would like to say that a flexible-chain macromolecule, getting into a limited space commensurable with its sizes, feels discomfort like we do. It is very visually shown in the course of a liquid chromatography. There is a separation of substances according to the sizes of their molecules, molecular weight and interaction to the surface of a sorbent along which there is a movement together with a solvent stream. When the macromolecule gets to a pore available to it, it stays there for some time and lags behind a stream. In the pores, the entropy of a molecule is lowered, as well as in the space between membranes (of which we have been just convinced). This reduction of entropy is increased with an increasing of the molecule size. This leads to the corresponding reduction of probability for the molecule to be in pores (i.e., a macromolecule "seeks to avoid falling into the pores").

On the other hand, the possibility of the adsorptive interaction with a surface of the pores, which manifests in an attraction and formation of temporary contacts, reduces the internal energy of molecules (because the energy of the attraction is negative). The value of the adsorptive interaction increases with the increasing molecular mass of macromolecules. A reduction of internal energy increases the probability of the penetration of the molecules into the pores. Adding up, these changes give a resultant increase or reduction of the free energy, which eventually defines the probability (and consequently, the frequency) of the penetration of macromolecules into the pores. The more often the molecule gets to the pores and the longer it is there, the more slowly it moves along a sorbent with a solvent stream.

According to this, if at a molecule's penetration into a pore the entropy interaction prevails over the enthalpy interaction (i.e., the entropy reduction is more than the reduction of the internal energy), then the molecules of bigger sizes move along a sorbent quicker than molecules of smaller sizes. Otherwise, the picture changes conversely. This can be easily seen in Figure 10, which shows the result of a thin-layer chromatography of polystyrenes on silica gel in a three-component solvent: cyclohexane-benzene-acetone. Varying the composition of the solvent, it is possible to change the value of the adsorptive interaction of the polystyrene macromolecules with an internal surface of silica gel. According to this, if at a penetration of a molecule into a pore the entropy interaction prevails over the enthalpy interaction (i.e., the entropy reduction is more than the reduction of the internal energy), then the molecules of bigger sizes move along a sorbent quicker than molecules of smaller value. In the drawing there are five plates with silica gel. Small amounts of samples of polystyrene with consecutive increase in their molecular weight were applied on each of them from left to right (on one line in the lower part). The plates were placed vertically in vessels with different compositions of solvent so that only the bottom edge was moistened. The solvent moved along the plates in capillaries between silica gel particles from below upwards, entraining the polystyrene macromolecules.

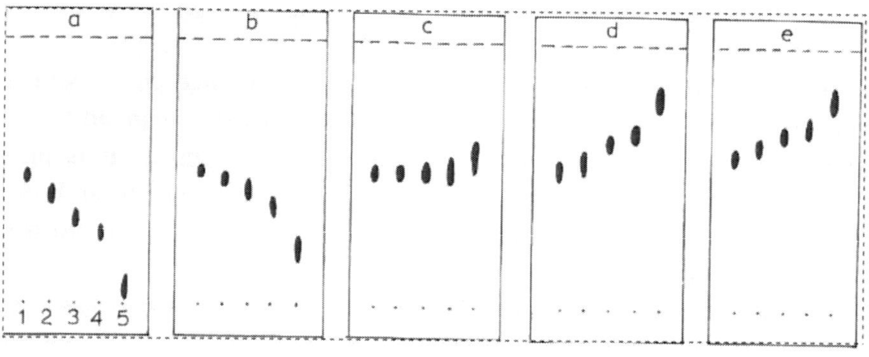

Figure 10. Thin-layer chromatography of polystyrene macromolecules using variable compositions of the solvent [37].

On the extreme left plate the adsorptive interaction exceeded the entropic interaction and small molecules moved much quicker than the large ones.

On the following plate the adsorptive interaction was reduced by a change of composition of the solvent, but still played a larger role than the entropic interaction. The speed of movement of the big molecules increased, though it remained smaller than for the small molecules.

On the third (central) plate enthalpy and entropy interactions completely compensated for each other and all of the molecules, irrespective of their molecular weight, moved with one speed. "The entropic discomfort" of the macromolecules at their penetration inside pores was dismissed by the adsorptive interaction leading to the loss of the internal energy.

On the fourth and fifth plates already the entropy interaction became prevailing and big molecules began to advance the small molecules at the movement along the plates.

The effect shown in the Figure [10] was discovered by the author with his colleagues in 1974 [37] and became the cornerstone of a so-called critical chromatography of macromolecules.

III. Correlative Changes of Entropy and Enthalpy As a Driving Factor of Natural Processes

Let us consider a few of the simplest examples.

1. Impact of Objects

Let us consider an object as an arbitrary system, and let external influences impact on this object. Let us say that as a result of the blow the considered object was deformed and its density became non-uniform. This means that the additional structure was imposed on its initial structure, and the entropy of the object decreased. Thus, if this object possesses heat conductivity, its temperature has increased so that the value of product sT has changed, not as considerably as it would have done if only an entropy reduction happened. The object has thus received some quantity of heat, i.e., its enthalpy has increased. To return to a condition close to the initial one, the acquired heat must be radiated into the surrounding space, and the deformation must be liquidated (at least partially).

The above process is the simplest example of mutually correlated changes in the entropy and enthalpy components of the free energy of the system under the action of external forces.

One can similarly consider the direct piezoelectric effect that is the effect of an anisotropic crystalline dielectric polarization under mechanical stresses, and the inverse piezoelectric effect, which is the emergence of mechanical deformations in the electric field [38]. In both cases, the original crystal symmetry is broken by the deformation of the electron shells and the corresponding displacement of atoms, i.e., in the crystal structure there appears disturbance, decreasing their entropy and enthalpy increases. Crystals come out of the thermodynamic equilibrium. According to Le Chatelier's Principle and the principle of the least action, processes against these changes start.

2. Direct Piezoelectric Effect

In the direct piezoelectric effect (as a result of the deformation of electron shells and of a shift of corresponding atoms) a difference of electrical potentials appears in the crystals and the associated electric field produces an electric current. After the passage of the current, mechanical deformations disappear and the crystal returns to equilibrium.

3. Reverse Piezoelectric Effect

In the reverse piezoelectric effect the disturbance caused by an electric field is damped by the corresponding mechanical deformations. Recently, a piezoelectric effect has begun to be used as a way of receiving cheap electric energy. For this purpose, crystal elements are placed under the asphalt road surfacing, and deform under the weight of passing cars. The difference of electric potentials arising in crystals is used for receiving electric energy.

4. Absence of Collapse of Atoms and Neutron Stars

The absence of collapse of nuclear nucleus (in the microcosm) and neutron stars (in the macrocosm) can serve as an impressive manifestation of the entropic repulsion. Without the entropic repulsion all matter would have turned into Black Holes as a result of a collapse.

5. Chemical Process of Dissolution of Substances

The chemical process of the dissolution of substances is described by equation (2). Entropic interaction and the subsequent tendency to increase entropy acts here as the driving force of the process, at the simultaneous minimization of Gibbs energy, G.

6. Osmosis Phenomenon

The phenomenon of osmosis and the associated osmotic pressure can also be viewed as a consequence of the entropic interaction by which the distribution of molecules in a solution on both sides of a semipermeable membrane aspires to be levelled and thus to raise the entropic value.

7. Gas Laws of Thermodynamics

Entropic interaction is a base of the gas laws of thermodynamics, substance transitions from one modular condition into another one and of phase transitions. It explains many regularities of statistical physics and is even considered to be the reason for gravitation. It is actively present in our daily practice.

8. Phase Transitions

The phase transitions or, as is less general, the transitions of substances from one modular condition into another, are used for receiving heat, its absorption and preservation. This possibility is caused by the change of entropy, s, during phase transitions and by the aspiration of the free energy of Gibbs, G, to remain invariable at the expense of the corresponding change of enthalpy, H.

9. Definition of Entropic Interactions

The introduction of the concept of entropic interaction looks natural to all natural phenomena. It should be considered as the quintessence of the

Universe and understood as the mutual influence of open thermodynamic systems on the state of each other by means of a change in their total entropy at a change in the entropy of one of them. The subsequent transition of all of the set of these systems to a more probable state is a consequence of entropic interaction. This transition is made through fundamental physical interactions - gravitational, electromagnetic, strong and weak - being affiliated from the entropic interaction and realizing it.

The entropic interaction is not a consequence of the existence of some entropic charge and a field accompanying it. It is impossible to say that the entropic interaction extends in space "from a point to a point" and transfers energy. Entropy is a countably additive function. Therefore, its change in any part of a thermodynamic system is at the same time a change of entropy of the whole system. Entropic interaction only reflects a certain "order" in space, "the state" of the space and of the physical systems which are in it, and, eventually, influences the power, behavior and evolution of these systems and of all the space as a whole, realizing itself through the four above-mentioned main interactions and related processes happening in the Universe, in the Solar system, the Terrestrial atmosphere, in Earth's atmosphere, inside the Earth and on its surface, and also through the corresponding biochemical reactions proceeding in plants and living organisms.

For a better understanding of the entropic interaction, we can use the following example. Imagine that you have a desire to perform some action, such as moving an object, lighting a fire in the fireplace, or planting a tree in the yard. And you're doing it. And the root cause of these actions is your desire to do them. Without desire there would be no action. The appearance of any desire leads to a certain structuring on relevant sections of your brain. The entropy of these areas (as well as of the whole brain and of your whole body) decreases. Your brain gets out of equilibrium, but it is eager to return to it. The return happens either when the desire is fulfilled or you cancel it. Here, as in a general case, the entropy change in any part of the thermodynamic system (caused, for example, by an external influence) is the root of the "turning on" of processes, seeking to maintain its status quo, to restore the previous value of the entropy in this part of the system, or at least the entropic change of the system in another location, so that the total value of the entropy would be closer to the original.

The formation of the Universe and the development of organic life on our planet will be considered later from positions of entropic interaction. And now, we will carry out a short digression to a process well familiar to us in the living Nature. Consider what happens when a trivial impact has taken place on a living tissue, such as an impact on the human body (which, in principle, can be regarded as a thermodynamic system). At the point of impact, the uniformity of the tissue on the body is broken, the tissue becomes "more structured"; on it, in addition to the old one (a kind of "over it"), a new structure - the "structure of an injury" - appears on the body. The entropy of the injured part of the body thus decreases as well as the entropy of the entire body as a whole (due to a countably additive nature of entropy). According to Le Chatelier's principle, processes begin to occur in the body (biochemical reactions), which oppose the entropic reduction associated with the structuring caused by the impact. As a result of these processes, a tumor appears, the temperature increases in the area of the injury, and gradually the "structure injury" begins to "melt." The organism spends a part of its energy (enthalpy) on these processes, the goal of which is to lead to the elimination of the consequences of the trauma associated with the impact, to lead to the healing of injuries. Ultimately, "the structure of injury" disappears, the entropy and temperature are restored, as well as the energy (due to the recharge from outside), i.e., the healing process takes place in full accordance with Eq. (2).

Any human disease can be considered from the point of view of the entropy perturbations; for example, a cold, producing a sore throat. An additional structuring, non-heterogeneity, appears in the throat. Entropy decreases, and the body spends energy, i.e., enthalpy, to stimulate the biochemical processes leading to the elimination of the additional structure: there is an abscess, fever and so on. The patient recovers, but becomes weaker (due to energy expenditure) and for the full recovery of the initial state there requires a certain energy "feeding" from outside.

More generally, we can say that all diseases are associated with a specific decrease in entropy caused by the appearance of some inhomogeneities of tissues, and their additional structuring (and subsequent inflammation). In the process of healing, these inhomogeneities and structuring are eliminated, and the entropy increases.

Proceeding from this, it is possible to expect the treatment of any disease if it is possible to turn on the mechanism of increasing the entropy of a sick body by means of external influence. The essence of this mechanism is in the activation of a protective reaction of the immune system. It seems that all immunotropic drugs work in this way, although their founders did not reflect at all on the entropic effect which is their cornerstone.

In summary, it is necessary to say that entropic interactions are a cornerstone of all natural processes happening both in the living and in the inanimate Nature. The subsequent sections are devoted to examples of their manifestation.

IV. Entropy and Time: Relationship between the Entropic Changes and the Rhythm of Time in the Natural Processes

"Entropy at a quicker rate... affords a more intense life process."
Erwin Schrodinger

As entropic value, along with a variety of other characteristics, defines the state of the chosen thermodynamic system, and a change of entropy sets the probability of transition of this system over time from one state into another, the question of the existence of a possible correlation between a course of time and an entropic change appears natural. The question of the possibility of such a correlation arose right after the introduction of the entropy concept into thermodynamics and statistical physics. As a prerequisite for that correlation served the single-orientation of time, expansion of the Universe and observed increase of entropy for all isolated thermodynamic systems over time. At the same time, again, the problem of the definition of the essence of time emerged, which was already exciting the ancient Greek philosophers. Personally, I like the definition of time given by the physicist and philosopher Ernst Mach [12].

"Time is an abstraction to which we come, watching changing objects of Nature." This short and extremely capacious definition of time, given by Ernst Mach, contains a profound meaning. First, it is argued that time is an abstract concept and not a certain entity. Secondly, it describes only a change in the state of natural objects.

To determine time, in daily practice there is the use of a so-called "astronomical time" (ephemeris time), which is measured by the number of revolutions of the Earth around the Sun. A shorter time interval relates to the pace of the Earth around its axis. And number 9 192 631 770 of periods of the electromagnetic radiation produced by the transition between the two hyperfine levels of the basic state of the cesium 133 atom defines a time interval of 1 second. (One year contains approximately $\pi\ 10^7$ *sec*).

Time is a way of evaluating and comparing processes occurring in Nature. When we talk about the duration of these processes, we are referring to that part of the revolution around the Sun that the Earth completed during these processes. And, as any process that occurs in a given system (an object or the Universe as a whole) by definition changes the state of the system, so time is a parameter, with the increase of which this change takes place. It is necessary to remind ourselves that the state of the system should be understood as a set of all of the possible characteristics inherent in the nature of the system. The absence of any changes will mean that time has stopped.

We will consider the behavior of entropy in thermodynamic systems. We will choose as one such system a spherically symmetrical extending gas cloud. Let us consider that it starts moving as a whole, in some chosen direction, continuing to extend, i.e., increasing the entropy. If we move together with the cloud in the same direction and with the same speed (i.e., being in one reference system together with the cloud), we will observe a symmetrical expansion of the cloud, i.e., it will be a sphere increasing in size. However, if we stop (this means we will occupy a position in the reference system that is not moving) and the cloud continues to move, we will observe the cloud as a deformed sphere, flattened from behind and extended from the front. And this deformation will increase all the time, as the sphere's back surface, facing us, moves in the direction from us with a speed of movement of the center of the sphere, reduced by the value of speed of the expansion of the sphere. The front surface of the sphere moves, accordingly, with a greater speed. The deformation of the cloud observed in our reference system leads to the reduction of entropy of the cloud. At the same time, the proceeding expansion of the cloud leads to the increase of its entropy. Algebraically summarizing, these entropy changes result in a smaller, i.e., slower, increase of the total entropy of the cloud.

We will notice that time behaves similarly in the Special Theory of Relativity, according to which time's rhythm in a tested system is slowed down with the growth of its speed.

If we place the gas cloud from our example into a gravitational field, the cloud will asymmetrically expand in the direction of the field's source. The expansion of the cloud will lead to the increase of its entropy, and the asymmetry will lead to some reduction of the entropy. Summarizing, these two effects will give a reduced increase of the cloud's entropy. In addition, during the expansion of the cloud, work against the forces of the gravitational attraction between cloud molecules has to be made. This effect, in addition to the two previous ones, will also slow down the cloud's expansion and, consequently, will also slow down the growth of its entropy. Furthermore, we will notice that time's rhythm in the General Theory of Relativity is slowed down in the presence of a gravitational field and at its increase as well. Thus, we come to the conclusion that the behavior of entropy is similar to the behavior of time in a formalism of the Theory of Relativity. Therefore, a correlation between the change of entropy and the time course for any thermodynamic system looks quite natural.

In Appendix 3 it is shown how this correlation can be calculated. That is, when there are some processes in a system, its state changes. This change can be described by means of a differential equation. Solving it under the corresponding conditions, it is possible to prove strictly that the change of entropy occurring thus is proportional to the interval of time during which it happened. Appendix 4 shows that it is possible to come to the same conclusion, proceeding from the principle of the least action. Moreover, it is also shown that at movement speeds which are approaching the speed of light, the decrease of entropy of a thermodynamic system can fully compensate its standard increase which is specific for the Universe. This means that the system's entropy will stop changing, all the processes in the system will stop, and the time will stop too. In particular, it explains the fact (postulated in the theory of relativity) that the speed of light is the maximal speed for any material objects.

On the basis of these results it is possible to formulate the following thesis:

Any process happening in an isolated thermodynamic system during the time of Δt changes the entropy of the system by the value proportional to an interval of time of $\Delta t = t_2 - t_1$, i.e., the changes of entropy Δs and time Δt are interconnected and proportional to each other. Also, the change of the entropy is positive, as the value $\Delta t = t_2 - t_1$ is always positive.

It is important to note that the cloud of molecules in our example can be replaced by just one molecule. Then, according to the Ergodic theorem [39] (proving that the average for an ensemble is equal to an average for time), it is possible not to follow a set of molecules in the cloud, but to consider only a single molecule. The probability of its being in any arbitrary point of the considered limited space will not differ from the renormalized appropriate density of molecules in the expanding cloud. This means that a random walk of this molecule will be such that this probability tends to be the same in any local area of the space.

It is also important that the external force acting on this chosen molecule (or on any other individual object) takes it out from the state of the random walk. The molecule starts to move in a certain direction. This means that, at the same time, the space surrounding the molecule becomes non-isotropic, which immediately reduces its entropy and its rhythm of time. When an external power is no longer valid, the molecule (the object) continues to move in the same direction (by inertia), increasing the likelihood of its presence in other points in space and thus increasing the entropy (both its own and that of the surrounding area). This explains Newton's first law of motion and the cause of inertia (i.e., tendency to increase entropy is realized by moving due to inertia).

Actually (as previously stated), the processes occurring in the local area of any open space are strongly dependent on the processes in the wider region as well as on the characteristics of the Universe, on processes occurring in it, and on the associated changes. The converse is also true, i.e., processes taking place in the local systems have some effects on the system of which they are a part.

The processes occurring in the Universe, affecting it as a whole, and the related speed of change of its state, define the rates of change of the entropy of the Universe and of the time flowing in it. It is expedient to refer to both of these rates as standard ones, i.e., for any concrete thermodynamic system there is a standard increase of entropy and a standard rhythm of time corresponding to it.

On the other hand, the state of this system can change under the influence of local external influences. In it there can be processes occurring in the background of processes that are general to all of the Universe. These local processes introduce some corrections to the change of the entropy of the system. These changes algebraically add up with the standard increase of entropy, setting thus its total local change. The rate of this total change of entropy defines an internal (individual) rhythm of time, characteristic for this system.

Thus, we come to a very important conclusion, which is that by changing (in one way or another) the entropy of a thermodynamic system, it is possible to regulate the rhythm of its internal time, and consequently also the speed of processes occurring in it, including the aging processes of objects from both living and inanimate Nature.

The notion of internal time is not new. For example, Newton identified two categories of time: absolute and relative. In his Philosophiæ Naturalis Principia Mathematica ("Mathematical Principles of Natural Philosophy") [40], we read: "Absolute, true, and mathematical time has no relation to anything external, it flows uniformly and is called duration ... Relative time is accurate, performed by ... means of a measure of the duration of a movement, it is used in everyday life instead of ... mathematical time."

That Newtonian absolute time is a kind of equivalent to the standard time introduced by us above. We can easily consider individual time to be Newton's relative time.

Einstein, in creating the theory of relativity, came to the conclusion that "in every inertial system there must be defined its special time" (the individual time in our definition) [41].

In a thermodynamic system, all processes slow down when heat is retracted from it (for example, the freezing of liquids) and accelerate with its tributaries the influx of heat (the melting of liquids). The decreasing of heat decreases the entropy of the system, and simultaneously increases it with the influx of heat. These changes in the entropy, algebraically adding to its standard growth, lead to its faster or slower growth, in correlation with the changes in the speed of the processes occurring in the system, i.e., in correlation with the changing rate of time in the system.

The Third Law of thermodynamics says that at the approach of the temperature of a thermodynamic system to absolute zero, the value of its entropy ceases to change. From the proportionality of change of the entropy and of the time, it follows that time at the absolute zero temperature also ceases to change, i.e., time stops; it does not exist. This situation was before "the Big Bang"!

Let us repeat once more that we know from the special theory of relativity that the rhythm of time depends on the choice of reference system. That is, each randomly selected system has its own internal rhythm of time that is different from the external rhythm of time, which is defined as the rhythm of time in a system of reference that is in a rest. The time is going more slowly with increasing speed. (I.e., a time interval between two identical events is less in the reference system moving with a greater speed. In particular, the time of the life of an unstable elementary particle in this case increases by the multiplier equal to the quotient of its total energy and the energy at rest.)

According to the general theory of relativity, time goes more slowly in a system with a greater mass and in the presence of a larger external gravitational field. This means that the internal time depends strongly on the choice of a reference system, its condition and location in space and, therefore, it is an integral part of the four-dimensional space-time continuum. Additionally, it is necessary to consider also a thermal balance with the environment, the system temperature, the value, direction and rhythm of the change of its entropy. In a thermodynamic system all processes are slowed down at the removal of heat from it (for example, while freezing liquids) and are accelerated with its inflow (while thawing liquids). At the same time as the heat removal, entropy of the considered system decreases, and with its inflow it increases, i.e., we have to understand that the state of the system and the rhythm of time in it are determined not only by the speed of the system and by the nature of its motion – inertial or non-inertial (as considered in the theory of relativity) – but also by the heat balance between the environment, the temperature of the system, the value, direction and the pace of change of its entropy.

However, it would be incorrect to think that entropy and time changes identically coincide. According to the principle of causality, the processes happening at an earlier time can have an impact on the processes occurring later, but the reverse is not true. As for the entropic changes connected with these processes, it is impossible to claim that for an isolated thermodynamic system the state with smaller entropy preceded a state with bigger entropy. Unlike time, entropic changes have a statistical nature. Each closed thermodynamic system (if it is brought out of an equilibrium state) seeks to go into a state of more equilibrium (i.e., a more probable state), though it can also go into a less probable state (i.e., one of less equilibrium). The entropy upon such a transition (when the system goes into a state of less equilibrium) will decrease, but time will continue to increase, without changing direction. This opportunity is very small, but nevertheless it exists and can be estimated according to Poincaré's recurrence theorem [42]. (The theorem states that certain systems will, after a sufficiently long but finite time, return to a state very close to their initial state.)

Nevertheless, for each thermodynamic system its time is a parameter that reflects the speed of change of its entropy. If, while comparing two systems, the entropy of one of them changes more quickly, it means that all of the processes proceeding in this system (including the time in the system) go quicker than in the other system.

I.e., it is possible to say the following:

Time may serve as a measure of entropy change, and entropy change is the quintessence of time.

V. Entropic Interaction as a Possible Explanation for the Negative Gravity at the Moment of the Big Bang

The section of physics called cosmology is engaged with the questions of the origin of the Universe and it researches the global phenomena occurring in it. According to the understanding developed in it, the Universe started after the "Big Bang" [5]. Why the Big Bang occurred (about 13.7 billion years ago), how the Universe was formed in its result and how it developed further, is explained by the cosmological "Inflation Theory" [4]. According to this, all of the World space was filled "by a physical vacuum" with a high density of energy. The condition of this vacuum was stable. In it there were no processes, i.e., there was no time. However, in each local area of this space (the size, commensurable with "Planck's length" $l_P = \sqrt{\frac{Gh}{2\pi c^3}} = 1.616\,10^{-35} m)$ casual quantum perturbations (fluctuations) of energy of the vacuum (not changing its stable state) took place. (Here G is the gravitational constant, h is Planck's constant, c is the velocity of light in a vacuum). But there were also fluctuations bigger than some critical value. And the vacuum "pushed out from itself" the material objects which possessed features of negative gravitation. Instead of an attractive interaction, these objects repelled each other. This was accompanied by an outcome of a huge amount of heat and proceeded during a very short period, considerably smaller than one second (rather close to the time of $t_P = \sqrt{\frac{Gh}{2\pi c^5}} = 5.39\,10^{-44} sec$, called Planck's time, during which light is passing in a vacuum of Planck's length. After Planck's time, the matter formed in this way and gained properties of gravitational matter, as we see it now. Its particles started to be attracted to each other by gravitation and correspondingly started to bend the geometry of the surrounding space, which was now filled by a vacuum with a low density of energy. As at the initial moment the sizes of this arising Universe were very small (an order of Planck's length), the gravitational attraction of the matter making it was very big, and the Universe had many chances for a collapse. However, the quantum tunnel effect allowed it to break the gravitational power barrier, on the other side of which it could continue to increase in size [43]. Thus, the masses and the sizes of the material objects formed in it began to increase from the sizes of elementary particles to the sizes of stars and galaxies, without breaking by their growth the law of energy conservation, because the energy of their gravitational attraction, being negative, compensated for the increase of their mass. These masses, which are running away extensively, received huge impulses at "the Big Bang". The appeared gravitational attraction was unable to stop this process and could only slow it down. Therefore, the galaxies formed in this way continued to run away with a small negative acceleration. However, about 4.5 trillion years later, there appeared in the vacuum a negative pressure surpassing the gravitational attraction, and the Universe began to expand with a positive acceleration that continues to the present day.

The emergence of negative pressure is connected with the distribution (which was established in the Universe) of so-called dark energy and dark matter, to which relate about 95% of all the mass of the Universe. Material objects (planets, stars, galaxies and their congestions) which we observe by means of our vision and devices increasing it (optical and radio telescopes) comprise less than 5% of the Universal mass. This was revealed by astronomers in the nineteen-thirties [44-46] and was convincingly confirmed in the 'sixties [47] on the basis of more modern equipment. The term "dark mass" reflects the fact that this mass does not interact with electromagnetic radiation, does not reflect and does not radiate it (at least, in the threshold available to our supervision). However, the "dark mass" actively participates in the gravitational interaction of objects. The conclusion about its existence was made on the basis of the stability of a conglomerate of galaxies rotating around their mutual center. The mass of these galaxies, measured by optical methods (Doppler effect), is essentially insufficient for holding them on the corresponding orbits at the speeds with which they move. The observed stability of these orbits requires a mass hundreds of times bigger. The ratio between the visible and the dark masses is usually illustrated as shown on the Figure 11.

Today it is supposed that the dark matter consists of briefly living elementary particles, penetrating freely enough into observed material objects. Scientists hope to discover products of decay of these particles in experiments at high energy.

The quantum fluctuation which brought about "the Big Bang" and the subsequent formation of our Universe could not be only singular. Similar fluctuations, being accompanied by a "Big Bang," had to happen also in other local areas of the Universe (but not in our Universe filled with low energy vacuum, as the energy is not sufficient for the implementation of "the Big Bang" as a consequence of the vacuum fluctuations). And there must be an infinite number of such areas. Respectively, there had to be formed (and there must continue to be formed) a lot of Universes, both similar to ours, and different from it. These Universes fill the World space, do not depend on each other and do not interact, i.e., they are closed systems that do not allow material objects to move from one such Universe to another. As there is an infinite number of them, there exists the probability of a multiple repetition of our Universe and of events occurring in it. For the same reason, the world's physical constants, characteristic of our Universe and allowing the emergence of organic life on our planet, were realized in a unique way from their possibly infinitely large set.

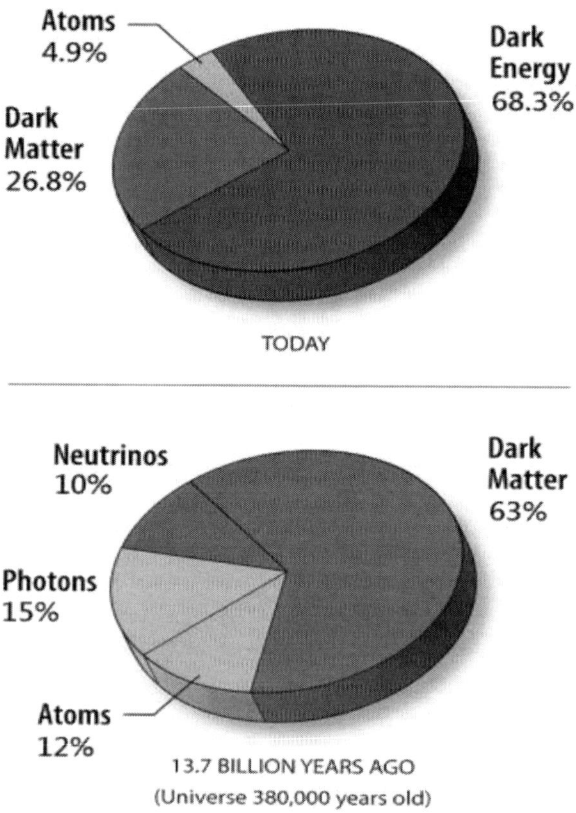

Figure 11. Distribution of the visible and dark masses in the Universe.

Such is, in general, the modern cosmological picture of the World, against which, apparently, there are no serious objections. The thermodynamic approach to natural phenomena allows us to add to the cosmological picture of the World as follows. We know from nuclear physics that the mass of an atomic nucleus is always less than the total mass of nucleons making it. This defect of mass is explained by “an expense” of it on the binding energy of nucleons in an atomic nucleus. The defect of the mass is reversible: the mass spent for binding energy does not disappear completely. If the atomic nucleus is to be dismantled to nucleons entering it, each of them will have the original initial mass. (The binding energy in an atomic nucleus is obliged by its emergence to strong and gravitational interactions of nucleons).

According to the General Theory of Relativity, the presence of objects (possessing mass) in space changes the space geometry, bends the space, and makes the space non-Euclidean. Take any handkerchief and stretch it by all four corners. It will give you some analogy of a flat Euclidean space. And now throw any massive ball at this stretched handkerchief. What will occur? The ball will additionally stretch the handkerchief, having pressed in it a pit, and will roll down to the handkerchief's middle, together with the pit. The geometry of the handkerchief changed. It became non-Euclidean. We received an analog of the bent space. And now we will throw a second ball at the handkerchief. It will press one more pit in the handkerchief, and then both of the balls will roll toward each other and will form a common pit on the handkerchief. So Einstein popularly explained the phenomenon of gravitation [48], i.e., the phenomenon of the gravitational attraction of objects, and the related curvature of space which becomes (conditionally speaking) a concave one.

“The expense” of mass on the binding energy, caused by their mutual attraction, reduces the gravitational curvature of space. And the reversibility of the defect of mass says that the interacting masses nevertheless continue to be present in space in full quantity and, therefore, still continue to bend it, making it concave, but to a somewhat lesser extent. Here it is high time to remember the entropic pushing away of interacting objects. The gravitational attraction between them generates the chosen direction in space, i.e., that the attraction additionally structures the space and thus reduces its entropy. The reduction of entropy is also affected by the rapprochement of objects occurring thus. According to Le Chatelier's Principle, Nature is inertial and opposes it, increasing the entropic pushing away of objects from each other. For this reason, the energy of the attraction is considered to be negative, in all calculations, i.e., it is basically equivalent to the negative mass ($0 > E = -mc^2$).

With distance reduction between masses, their gravitational interaction increases, the binding energy increases, the entropy decreases, the defect of mass becomes greater, and it is possible to imagine an extreme case where the defect of mass is equal to the mass of the system, i.e., the matter possessing mass appears as though in some kind of virtual situation. The space will thus look massless, filled only with a vacuum with a definite density of energy.

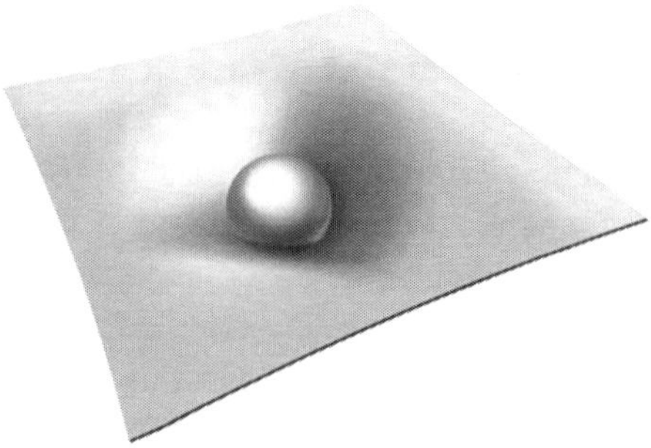

Figure 12. Model with which Einstein popularly explained gravity.

The World space before the Big Bang was such. It was flat (Euclidean) and was filled with vacuum consisting of very dense unstructured "chaotic" matter (with a density of $\sim 10^{97}$ kg/м^3), being in a virtual state. As a result of local quantum fluctuations, the vacuum density could increase in its small area above some critical size. Respectively, the entropy decreased. There was a structuring of this matter which captured (not at once) all the area where the fluctuation took place, but expanded on it gradually, yet with a very high speed. We observe something similar at the freezing of any liquid. When the liquid is cooled and is structured, its entropy is lowered, crystals of ice are formed and gradually fill up all of the liquid volume. At the same time, the liquid radiates some amount of heat.

Elementary particles were formed when the space was structured. They began to bend it gravitationally and to interact in such a way that all four fundamental interactions known to us acted as a unit, bending space in the opposite direction. For a short time (close to Planck time) this curvature prevailed, i.e., the defect of mass of the formed particles was greater than their total mass, which led to their mutual pushing away. The structuring of the World space continued after the Big Bang. The family of elementary particles grew. They united, forming atoms, molecules, high-molecular compounds and so on. The entropy of space connected with this structuring decreased as appropriate, but expansion of this structured space led to the entropy growth. Summarizing, these two opposite tendencies gave the total increase in entropy that occurs now.

According to the cosmological theory of inflation, it is possible to assume that such a scenario is implemented only in a highly energetic vacuum and cannot take place in our Universe filled with vacuum with a low density of energy. The energy of the vacuum of our Universe is what remained from the energy of the initial highly energetic vacuum after "the Big Bang," and it isn't enough for a subsequent creation of a "Big Bang." However, when in any local area of our Universe the density of matter reaches some extremely big value, the matter can collapse, forming "Black Holes" [49].

Figure 13. Simulated view of a Black Hole in front of the Large Magellanic Cloud.

A Black Hole is an area in space whose gravitational attraction is so great that no material objects, including light quanta, can leave it. According to modern understanding, there are four scenarios for the formation of a Black Hole:

1. Gravitational collapse (catastrophic compression) of a rather massive star at the final stage of its evolution.
2. Collapse of the central part of a galaxy. For example, in the center of our Galaxy there is a Black Hole known as Sagittarius A.
3. Formation of Black Holes at the time right after the Big Bang as a result of fluctuations of matter.
4. Emergence of Black Holes in nuclear reactions of high energy.

Each of these scenarios connects the formation of Black Holes with a significant growth in gravitational interaction and the corresponding reduction of the entropy, caused by the local increase in the density of matter in real physical systems (objects). However, one more essentially different scenario (the fifth), is possible. Local quantum fluctuations of density of vacuum of our Universe, naturally, can lead both to an increase in its density, and to a fall in it. In both cases the entropy of space in the area of fluctuation will go down, the space will be structured and bent. But if an increase of the density makes the space "convex," then the fall of the density makes the space "concave." The material objects possessing mass, when they get into a "convex" space area, will seek to increase their entropy. For this purpose they will seek to leave this area, as happened during the Big Bang. And the objects that get into a "concave" area will seek to increase their entropy, filling that area, i.e., the "concave" area of space, formed in the manner described above, falls under the definition of the Black Hole, given above.

The possibility of a multiple repetition of the situation that led to the Big Bang does not contradict the described thermodynamic approach. However, it is also possible to allow other scenarios. For example, the structuring of space after each Big Bang does not lead to the formation of a new closed Universe, but supplements the previous one. This addition each time can differ by another set of world constants. Casual realization of the constants can give such characteristics as those that we have in the observed part of our Universe which is unique and fills all the World space.

The scenario according to which this unique Universe was formed after only one Big Bang is also possible. Thus structuring process (in the process of the expenditure of energy of vacuum) expanded (and continues to expand) over all the World space (as crystals of ice fill liquid which is freezing and is giving heat). Material structures possessing a casual set of world constants are formed in various parts of the World space.

It makes sense to emphasize once again that the entropic pushing away of material objects leads to an increase in the entropy and enthalpy of the system of these objects, and in a small time vicinity of "the Big Bang" led to a "negative" curvature of physical space, having made it convex for a short time.

The entropic repulsion of material objects continues to act also when they have moved away from each other at a significant distance. That repulsion transforms the totality of these objects into a more probable state and realizes itself by means of a "negative curvature" of space. In the scale of the Universe, the negative pressure generated by the specific distribution of the dark mass is responsible for this curvature. The dark mass strives to a more probable distribution in space, and its interaction with objects having "visible" mass can be conditionally considered analogous to interactions which lead to the widening of a gas cloud. In each of these cases the behavior of observed objects (galaxies, molecules of gas) is determined by their interaction with invisible subparticles. As a consequence of this, a gas cloud expands, galaxies continue to run away (with some acceleration), and the Universe continues to expand.

The state of weightlessness is some analogy of a situation in which the defect of the mass is equal to the mass itself. This effect has a place at a free fall of objects or at movement along a curvilinear trajectory with a corresponding speed; for example, on an orbit of an artificial satellite of Earth. In the latter case, when the speed of movement of the object exceeds the first cosmic speed (8000 km/s), "the defect of the mass" becomes more than the mass itself, and the object leaves the field of the terrestrial inclination (just as the virtual matter was realized from the original (initial) vacuum at the Big Bang).

The evident analogy of the Big Bang can also be represented conditionally, as follows. Imagine an iceberg floating on an ocean. Its largest part is hidden under water; it is not available to our sight and in relation to it, it is as though in a virtual state. Now imagine that the water density sharply increased by several times. What will occur? There will be an entropic repulsion! The system (water-iceberg) will seek to increase the entropy and to pass into a state of more equilibrium. The water will push out the iceberg. It will fly up over the water surface and only when its speed is reduced by an attraction to the Earth will the iceberg start falling down. It is possible to say that it is in a similar way that the highly energetic vacuum "pushed out" of itself matter after one of its quantum fluctuations and gave life to our Universe.

Using this analogy for an explanation of the formation of a Black Hole (according to the fifth scenario) assumes a fall of density of the water surrounding the iceberg. Thus, the iceberg will sink, forming a funnel imitating a Black Hole.

Figure 14. Iceberg.

VI. The Origin of Organic Life on Earth

"What is the characteristic feature of life?
When is a piece of matter said to be alive?
When it goes on "doing something," moving,
exchanging material with its environment."

Erwin Schrodinger

The appearance of organic life on the Earth may seem surprising if one takes into account only the entropy changes associated with it. Indeed, on the one hand, the entropy (at least in our region of the Universe) is continuously increasing with time, reflecting the Universe's tendency to homogeneity and its disordering. On the other hand, the appearance of organic life is the result of a sharp local ordering and the corresponding decrease in entropy. It would seem that we are faced with two conflicting conclusions. To avoid this, Prigogine [17] proposed considering the entropy of any system as the sum of its local and external components. At the same time, to perform the second law of thermodynamics, the increase of the external component should always cover the reduction in its local part.

In addition, this contradiction is resolved if we recall that the entropy alone does not dictate the development of the world. In accordance with the principle of least action, the world, as a thermodynamic system, tends to remain in the position of thermodynamic equilibrium, or to strive for it, if it is taken out of this state. The Gibbs free energy, G, is one characteristic that is responsible for this state and its evolution over time. At the equilibrium, the value of G is minimal. Entropy, S, is only part of the Gibbs energy G = H-Ts. Therefore, if, as a result of some evolutionary process, the entropy decreases, the enthalpy part, H, of the free energy must also be reduced, otherwise the probability of this process will be negligible.

It can be assumed that the original minerals, consisting of individual chemical elements, were randomly distributed in the earth and on its surface. Over time, they could dissolve in water and other liquids, and then interact with each other to form molecules of more complex substances. Finally, this process leads to the formation of macromolecules, i.e., polymers, among which there are amino acids and nucleotides. At the following stages of evolution the amino acids unite in a polymeric chain with a peptide bond, forming proteins, and the nucleotides form deoxyribonucleic (DNA) and ribonucleic (RNA) acids.

Subsequent evolution leads to a more structured state of matter, to the formation of living cells. But at every step of the structuring, the formation of new chemical bonds is energetically more favorable (the Gibbs free energy, G, decreases) than the separate existence of the connected elements.

The transition from a unicellular to a multicellular organism is accompanied by an increase in entropy. Perfecting the multicellular organism leads, on the one hand, to its additional structuring and to the reduction of its entropy. On the other hand, it leads to the growth of the entropy due to the increase in the size of the organism. For example, the brain of Homo sapiens is much more structured than the brain of a primitive man, but at the same time the brain of Homo sapiens is much larger, especially if we compare not the volumes and masses of the brains, but their surfaces.

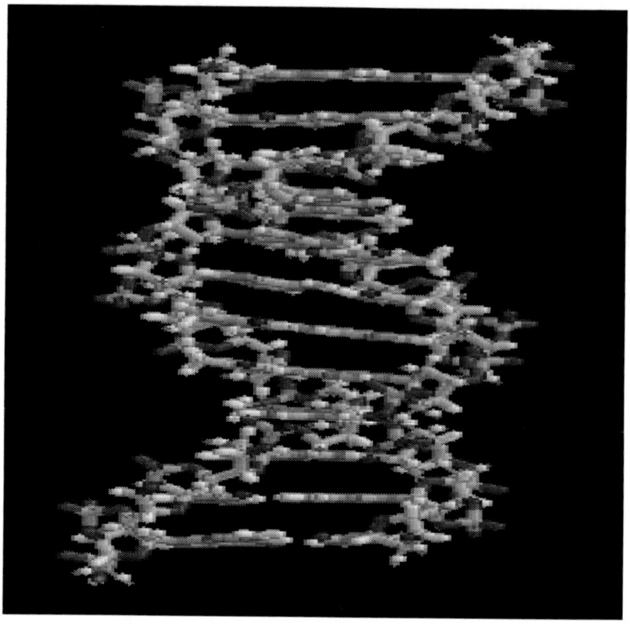

Figure 15. Double spiral of DNA.

It should be noted that each process, each event, everything that happens in the Nature, is followed by an increase in entropy. It also belongs to living organisms. They continuously increase their entropy and thus approach a state with maximum entropy, which is equivalent to death, i.e., a lack of any processes. To prolong their life, they have to interact constantly with the environment, lowering entropy and thus maintaining orderliness. Living organisms reach it, in particular, by means of a metabolism, i.e., the complex of physical, chemical, and biochemical processes occurring within a living cell or organism that are necessary for the maintenance of life. In metabolism, some substances are broken down to yield energy for vital processes, while other substances, necessary for life, are synthesized. Breathing, the consumption of food and drink, and the return to the environment of the waste products of this activity accompany these processes. Air, food, and water represent the kinds of structured (ordered) matter with the entropy lowered in comparison with the individual molecules making them up. Getting into a living organism, they lower its entropy, detaining its standard growth. In the course of a metabolism these products are destructuring, and their derivatives unite in new structures. Thus, the total entropy of an organism continues to grow due to the destruction of old formations and the emergence of new ones.

I.e., in a living organism there is continuously "a new order," thanks to the consumption of "some order" from the environment. In other words, the concept of a metabolism with "order from order" works here. This was suggested in 1944 on the basis of thermodynamics by the author of the basic equation of quantum mechanics, E. Schrödinger, in his book "What is Life?" [50].

It is important to understand that the new formations, appearing in an organism as a result of metabolism, act in a dual way. On the one hand, they decrease the entropy, but on the other they increase it because of their quantity growth. The total entropy change, connected with metabolism, turns out to be positive, although it is less than it would be in the absence of metabolism.

The use of the thermodynamic approach allowed Jeremy England and his colleagues to develop an alternative to the Darwin theory of evolution. In their works [51] it is shown that the more rapid growth of entropy of living organisms can be the defining factor for the survival of biological objects in their competitive fight for existence. This result correlates very well with the idea stated by E. Schrödinger in the above-mentioned book and quoted earlier in the epigraph to Paragraph IV: "Entropy at a quicker rate… affords a more intense life process" [50].

Here it is pertinent to remind ourselves that with entropy growth the thermodynamic system passes into a more probable state. Entropy grows quicker in the world of animate Nature than in the inanimate world. This means that the tendency to transition into a more probable state is more notable in the animate world. A consequence of this is preferable evolution of the Universe, at which islands of the living (organic) world arise and expand in the Universe. And the Future belongs to this World. From here, it is easy to come to the conclusion that a set of a hundred world constants (the mass of elementary particles, the characteristics of fundamental interactions, the value of the velocity of light, etc.), realized as a result of the evolution of the Universe and, in particular, on our planet, are not simply casual, but are natural. Theoretical calculations and computer simulations prove that the insignificant deviation of the value of at least one of the constants would have made the emergence of organic life impossible.

The DNA structure uniquely determines the genetic characteristics of a living organism. Slight changes (e.g., a shift of only two adjacent nucleotides) may lead to a very significant genetic change. Such a significant change in properties caused by a trivial change in the order of the macromolecular chain is appropriate not only for DNA, RNA or proteins. Relatively simple molecules, called optical isomers (or enantiomers) [52], dramatically change their properties because of a simple rearrangement of only two of their atoms. One of these isomers interacts properly with the appropriate protein and can be used as a medical drug, and the other does not possess this quality. Separation of the optical isomers can only be made chromatographically, using special so-called chiral sorbents [53]. And all this takes place despite the fact that the optical isomers consist of the same atoms and have the same chemical formula.

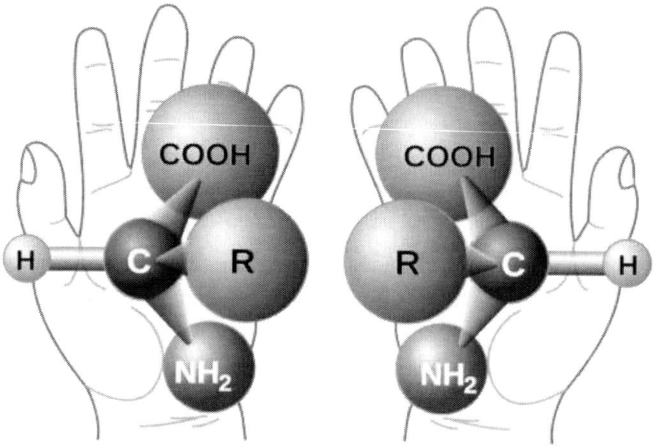

Figure 16. Two enantiomers of amino acid.

At first sight, such properties of optical isomers look quite strange. However, a very simple explanation exists for them. To work as a medical drug, the molecule of an optical isomer has to be built into a special "pocket" of a certain protein molecule; i.e., the small drug molecule has to enter into this "pocket," as a key into a keyhole. And for this purpose its form and size have to "be strictly verified." A shift of at least one atom will prevent the optical isomer from entering the protein and being used as "medicine." A simple explanation, isn't that so? It seems that in Nature not everything is as complicated as it seems at first sight.

This example once again demonstrates how the "entropic order governs the world." Particularly remarkable is the fact that our mortal body is "governed" by the immortal order of DNA. Today it is possible to determine through DNA analysis who our ancestor was: King Solomon, Alexander the Great or Ivan the Terrible. Let millennia pass, let many generations of people change, but the DNA of our remains will stay unchanged. Among other things, this will be our non-vanishing trace in life.

Chapter 2

Application of the Thermodynamic Approach to the Processes Happening in the Human Organism

I. Origins from Childhood

I distinctly remember the following event from my preschool childhood. I played with a group of classmates in the backyard of a large house in which we lived. In the same yard was a small house where the family of a janitor lived. This family from time to time had a visiting relative - an imposing middle-aged man with a stylish hat and a cane. The children knew that he was a hypnotist and each time asked him to demonstrate something from his theatrical repertoire. I remember when he invited us to join hands with our fingers interlaced, and we were not able to separate them without his touch. Once, he put some kids to sleep and went for a moment into the janitor's house. It so happened that, in his brief absence, the mother of one of the hypnotized boys came into the backyard and was terrified to see the children sleeping on the playground and not responding to her attempts to wake them up. The children who had not been hypnotized (including myself) began to reassure her, and someone knocked on the door of the janitor's house. The hypnotist came out, apologized, touched each of the sleeping children, and they awoke. The incident was over, but it always remained in my memory as a mystery requiring explanation.

A few years later I met another hypnotist. It was in a scout camp. Our counselor (his name was Misha) had learned the art of hypnosis. We somehow found out about it and often asked him to demonstrate his abilities. I remember kids who had been hypnotized by Misha, tearing leaves from the surrounding trees, thinking they were apples, or tearing and munching on grass, believing they were gathering berries.

Similarly to the events in the courtyard of our house, these hypnotic sessions left indelible emotional impressions. I became the person who constantly wanted to know "why?", tormenting parents and their friends. These questions required explanations of the hypnotists' abilities. Why, when hypnotized, could we not separate our hands? Why didn't the children, lulled to sleep, respond to attempts to wake them up? Why did the leaves on the trees seem to be apples to the hypnotized kids? And so on. I didn't have any intelligible answers to these numerous questions of "why", and they became a constant irritation on my consciousness.

II. Experiments with "Magnificent Seven" (People Having a "miraculous power," Possessing Paranormal and Healing Abilities)

Do you remember the Japanese movie "Seven Samurais" and the American movie "The Magnificent Seven"? In both of these films, seven top-class brave professionals saved defenseless villages from robbery and violence. And I'm grateful for the appearance in my life, too, of seven outstanding professionals who have each played a key role in developing the "thermodynamic approach" that led to the writing of this book. These professionals have slightly opened the hidden secrets of Nature for us; they have forced some seemingly incurable diseases to retreat, saving many lives. Three of them continue their noble mission to this day. And may God grant them good health and continued success! I will try as best I can to introduce them to you and explain their abilities. I will introduce to you the first five of my "Magnificent Seven" immediately, in the chronological order of my acquaintance with them. The final two of the "Magnificent Seven" (Tatiana Trofimowa and a Brazilian medium known as John of God), I will discuss later in the context of a theoretical explanation of their phenomenal abilities.

Number 1 - Wolf Messing: Hypnotist, Seer, Human-Legend

Figure 17. Wolf Messing.

-Seven or eight years after my first meeting with the phenomenon of hypnosis, which so shook my childish imagination, I had a memorable encounter with a man known throughout the world for his unusual, exceptional psychic abilities. It was Wolf Messing. It turned out that a niece of Messing's wife (a few years younger than me) studied in our school. He was persuaded to speak to the students at our school. He recounted some interesting episodes from his life, guessed the words written on the blackboard behind him, and then went down the hall wearing a black blindfold, inviting everyone to hide anything, and explaining that he would try to find what had been hidden while he was blindfolded. I must say that the show was a little creepy: an elderly man of extraordinary appearance with a wide black blindfold on his eyes, walking between rows, guided by the hand by one of our high school students. From time to time he stopped and told us what had been hidden, who had hidden it and where it had been hidden. I also decided to participate in this. One of the buttons on my sleeve was loose. I tore it off and put it in a shirt pocket and raised my hand as a sign of willingness to participate in the experiment.

Guided by the hand, Messing stopped a foot away from me (I was sitting by an aisle) and asked me to think about the object that I had hidden. He asked me to mentally imagine this object and the place where it was hidden. Then he paused for a few seconds. All of the people in the hall held their breath. Messing, without touching me and while remaining blindfolded, slowly began to speak, giving a short pause after each sentence. He said that I had hidden a small white button in my upper right shirt pocket and that the button had been previously apparently sewn to the shirt. *He guessed absolutely everything!*

Number 2 - Michael Kuni: Champion of Arithmetic Calculations, Psychic, Hypnotist

Figure 18. Michael Kuni.

Another seven years passed. I was already a 4th year student in the Physics Department of the Leningrad State University, when one day I saw an advertisement in the nearest club for a forthcoming meeting with Michael Kuni - widely known for his amazing psychic and telepathic abilities. Of course, I went to this meeting and even joined in a group of volunteers, whose task was to control any possible fraud in the ongoing demonstrations by Kuni. I was on stage, near Kuni, and had to check the quality of the bandages on his eyes; I also took part in the selection of participants for experiments, etc. Kuni stunned everyone by drawing with charcoal (on sheets of white paper) the portraits of two spectators whom we had selected; he had not seen them, since his eyes were covered with bandages, impermeable to the light. After he had taken off the bandage, a large map of the USSR was brought in, on which the Maestro showed the places of our births, accompanied by the respective dates. Then I was asked to write several multi-digit numbers (5, 7 or 10 - I do not remember exactly) in arbitrary places on a blackboard. The board was installed at a distance of several meters behind Kuni and could be rotated. While I was writing these numbers, Kuni was talking to the audience. Then, at his command, one of the volunteers began to rotate the board fast. He turned to it, contemplating it for a few seconds, then turned away and wrote the sum on the other board. After this, he again began to converse with the audience, while one volunteer from our group wrote all of the numbers in a column to total them. The result coincided with the amount previously written by Kuni. The presentation lasted more than an hour and there were many other demonstrations. At the end of each one the audience burst into applause.

I should say that Kuni not only performed in front of the audience, but also actively participated in scientific experiments conducted by scientists of different professions. The physicists of the Joint Institute for Nuclear Research in Dubna, which included the world-renowned academicians I. Kurchatov, V. Veksler, and B. Pontecorvo, wrote in their report that experiments with Kuni demonstrated enormous untapped reserves of the human brain, which in interaction with the environment is able to perform wonders. This has been confirmed by experiments carried out by physiologists, and scientists from the Bechterev Research Institute even claimed that their findings might be physiologically substantiated. However, to date we are still awaiting such studies.

Number 3 - Nina Kulagina: Psychic and Telekinetic

Turning to the next meeting, which may have been the most significant meeting for me, I want to emphasize that all of them occurred spontaneously and seemingly without any deliberate initiation on my part. Fate, or some unknown law related to my desire to get answers to the "why"s that seemed to itch me, led me to this chosen path.

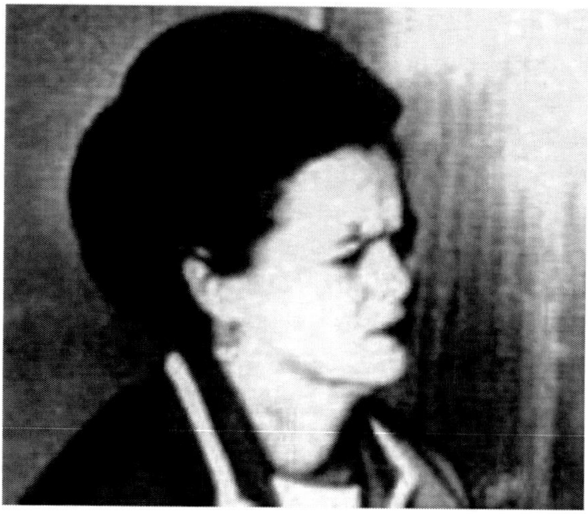

Figure 19. Nina Kulagina.

One spring day in 1968 I came out of the Public Library of Nevsky Prospekt, and at the House of Scientific and Technical Propaganda saw an advertisement for a lecture devoted to the phenomena of dowsing, telepathy and telekinesis.

The lecture was due to begin in 15 minutes. I had time to attend, so I went up to the second floor to the auditorium. There were a lot of people (maybe a few hundred). As it turned out, most of them were geologists. Many of the people in this profession were trying to practice the ancient method of finding groundwater and various minerals with the help of a twig of a vine [54]. It is a flexible twig (approximately a meter long), cut from a tree and stripped of leaves and boughs. It is held by its ends and bent into an arc, while the operator slowly moves around the area. It is assumed that in the presence of underground water or minerals, the vine twig will spontaneously begin to rotate in the hands of the person holding it. The relevant literature describes successful cases of using this searching method.

In the present day, enthusiasts have replaced the vine twig with a rod that is a rectangular metal wire frame with one or two handles. I have been told of searching with such frameworks, even with the use of a helicopter, but I have not seen any documented positive results. Interest in dowsing is especially typical for geologists, which explains their massive presence at this lecture.

Figure 20. Dowser (XVI century engraving).

For reference: telepathy is understood as a hypothetical brain's ability to transmit thoughts, feelings and images to another brain or organism at a distance or to take them from it without using any known means of communication.

Telekinesis is the ability of a person's effort of thought to make an impact on physical objects (in particular, to move them).

The lecture began with a speech from a mathematician of the Leningrad State University, associate professor Sergeyev. Its first part was devoted to phenomena of telepathy and telekinesis. According to Sergeyev, these can be explained by means of casual processes connecting interacting individuals and objects surrounding them. He wrote a number of equations on a board, and as an example of their use began to tell us about his experiences with Nina Sergeevna Kulagina.

At that time, I was interested in the application of the generalized stochastic processes to the problems of quantum field theory, and I was familiar with the mathematical apparatus used by Mr. Sergeyev. From my point of view, in his calculations, essential inaccuracies were allowed, and beside this, I had very negative information about Kulagina.

At the end of the nineteen-fifties, L. L. Vasiliev worked at Leningrad University as a professor of psychophysiology. His book "Mysterious phenomena of the human mentality" enjoyed a certain popularity. He organized a lab, conducted research in the field of parapsychology, and spent some time experimenting with Kulagina. In the mid-sixties I incidentally became acquainted (in one of the halls of the library of the Academy of Sciences) with several former employees of Professor Vasiliev, who were participating in these experiments.

From them I learned about these experiments, and also discovered that none of their results had been published because Professor Vasiliev was not sure about Mrs. Kulagina's honesty, and thus about the reliability of the data obtained. I was told that there were always problems with repeated experiments, that Mrs. Kulagina had at times been caught trying to eavesdrop or spy on the information which she had to receive telepathically. Knowing all this, and due to my youth, I could not remain quiet and at the end of Mr. Sergeyev's performance, I dashed onto the stage and gave this information out to the audience. The audience shouted, digesting it, and during an interval I was attacked by a furious woman, who appeared to be between 45 and 50 years old. The woman was Mrs. Kulagina.

"You don't know me," she exclaimed, "and you fill people's heads with foolish and incorrect information, when you speak about my remote past. Yes, at that time not everything turned out for me. I knew I was being watched. But I really wanted to convince Professor Vasiliev about my paranormal abilities, which at the time I was just beginning to realize. And it was very difficult to show them in a stressful situation where you are being watched by a dozen pairs of disbelieving eyes. But now, I have significantly more experience; I can and am ready to show it to you." With these words, she dictated her home phone number to me. We were already surrounded by the people in the audience, with some of them asking Mrs. Kulagina to give them her phone number too. But she flatly refused, saying that she would demonstrate her paranormal abilities "only to this mistrustful young man."

As far as I could, I prepared myself for the experiment with Mrs. Kulagina. For experiments on telekinesis, I stocked myself up with a glass cap on which electroconductive transparent paint had been sprayed. For telepathy and for control of the absence of possible hypnotic suggestion I took with me one more physicist, Misha Tennikov, who was extremely faithless regarding the existence of paranormal phenomena.

After talking on the phone, we arrived at the apartment of Mrs. Kulagina, who lived near the Avtovo metro station. She was waiting for us with her spouse (a recently retired former engineer).

I asked Mrs. Kulagina to begin with the simplest experiments on telekinesis. She agreed, put a small compass on a table, concentrated, and began to move her palm over it at a distance of about 20-30 cm. Her spouse explained to us that this was how she adjusted and entered into a special state. Mrs. Kulagina couldn't explain the essence of this state that she thus felt. From experience, she only knew that her paranormal ability turned on when the arrow of a compass started rotating, following the rotation of her palm. This took place in approximately one minute. The arrow, as if attached, rotated together with her palm clockwise and against it. Several times I ran a hand between the compass and Mrs. Kulagina's hand, to be convinced of the absence of any threads invisible to us. Misha looked under the table in search of a possible magnet, closely watching the spouse, who was sitting at the same table between us and Mrs. Kulagina (Misha and I sat opposite her). Mrs. Kulagina was dressed in a dress with short sleeves that excluded the presence of a magnet hidden on her body.

At the lecture where I first became acquainted with Mrs. Kulagina, Mr. Sergeyev said that she could move light objects through her willpower; in particular, an empty matchbox. Misha, a heavy smoker, took one from a pocket, emptied it of matches and put it on the table, short side down. Mrs. Kulagina brought her hands over the box and began to move them in different directions. The box smoothly followed the movements of her hands, seemingly without the friction of the table surface, as though it were on the thinnest pillow of air.

I changed the setting of the experimental statement a little bit: I put a sheet of white typewriting paper (which I had brought with me) on the table. Then I put the box on the sheet, drew a pencil line around it, and asked Mrs. Kulagina to move it again. She did it without any apparent effort. When the box was far from the original delineations, I asked her to stop it in this position and again traced around it with the pencil. In doing so, I wanted to make sure we were not exposed to hypnotic suggestion, and that the box had actually been moving.

I put this sheet into my briefcase, to later confirm the reality of the box's movement. Then I put a metal sheet (which I had also brought with us) on the table. I covered it with a sheet of white paper, put the box on it, and covered the box with a glass cap, which we had also brought. After that, I grounded the cap on the clean surface of a steam heating pipe and asked Mrs. Kulagina to repeat the experiment with the box under these conditions. Referring to the phenomenon of electromagnetic induction, I told Kulagina about what she had demonstrated to us. I explained that the human body is filled with a variety of different ions, i.e., charged particles. When she strained, some (considerable) amount of these ions rushed to her palms. Being charged, they induced charges of opposite signs on objects that she exposed to these effects (in this case, a matchbox). The Coulomb attraction between the palms and these items makes the items move, following the palms. The electrically grounded cap, imitating the famous (for physicists) "Faraday box," would not allow the electric field of her palms to penetrate and induced charges on the surface of the box, so it could not be "telekinetically" moved.

My explanation very much disturbed Mrs. Kulagina's spouse. He told us that once she doubted the achievement of a desirable result, it became very difficult for her to achieve it. Nonetheless, she started the experiment.

A few minutes later, she had no result. She reddened, and told us that she had started "becoming deaf" (this was her expression). This disturbed the spouse even more. He told us that when her efforts became arduous, her pulse and pressure tended to jump up so sharply that it sometimes became necessary to call for an ambulance.

In our case, this did not occur. Within five minutes the box shook and, to my astonishment, smoothly moved a little. The explanation of the "Kulagina effect" I had given did not work. Meanwhile, Mrs. Kulagina's face continued to remain very intense and red. Seeing that this experiment was taking a toll on her. I asked her to stop it. I lifted the cap and traced around the box with a pencil again.

The telekinetic experiment planned for this occasion was finished. Mrs. Kulagina began to tell us about some offenses connected with her recent performances before various groups of scientists. In particular, she had performed before physicists of the Ioffe Physical and Technical Institute of the Academy of Sciences of the USSR.

The members of the expert committee had been shocked by what they saw, and applauded, but when it came time to sign the protocol, they suddenly became cautious and asked her to repeat the experiment for the second time in a week, so Mrs. Kulagina told us. She agreed to repeat the experiment. Again, it all turned out successfully. In these experiments, among other things, she guessed symbols that the committee members had drawn with a pencil on paper. Mrs. Kulagina was in the next room, overseen by an observer. The probability of her guessing correctly was very high (I do not remember the numbers). One of the committee members then suggested that Mrs. Kulagina had a very fine ear that allowed her to hear the creaking of the pencil on the paper, and to determine what symbols they were drawing by the creaking. She was told about this and it was "explained" to her that the pencil would be replaced by a very "soft" ballpoint pen, that there would be no creaking and that she would fail to "guess" the symbols. This was done, and the percentage of correct guesses declined. The commission members were relieved and signed a protocol with the appropriate conclusion, specifying that it was possible that not everything had been correct in other experiments. It is curious that forty years later, I met a wonderful physician-psychiatrist in Boston (US) called Tonkonogy Iosif Mikhailovich, who had been one of the members of this commission. Our conversation accidentally touched upon telepathy and Iosif Mikhailovich remembered Mrs. Kulagina's experiment and confirmed her story. He remained convinced that the idea of the creaking pencil was true and that therefore Mrs. Kulagina had not demonstrated anything "supernatural." I suggested that after replacing the pencil with a ballpoint pen she had lost her confidence in her ability and, as a consequence, the percentage of correctly guessed symbols had decreased, to which Iosif Mikhailovich reacted skeptically.

Misha and I continued to talk to Mrs. Kulagina, and she offered to show us something besides telekinesis. She stretched her palms toward us and focused our attention on the traces of burns that were clearly visible on them. She explained that she could cause a burn to appear on someone's body by touching it, and that she would also receive a burn at the same time. With our consent, she clasped my left wrist and Misha's right one in her hands. A few minutes later, Misha felt a burning sensation and asked her to let his hand go. We saw a red spot, showing that the burning sensation was real. I also felt a burning sensation, but asked Mrs. Kulagina not to take her hand off me for some time. She did not mind, but warned that she could cause burns of any degree. The burning sensation on my wrist intensified and after a few minutes I stopped this experiment. The red spot on my arm remained for a few weeks.)

Mrs. Kulagina offered to conduct another experiment for us. She said that she could substantially raise the heartbeat of any person. One would only need to be with her and watch her eyes for the heartbeat to rise. I agreed to do this. We sat opposite each other at a distance of about two meters, and looked into each other's eyes. Misha measured my pulse with his watch and every minute he read the numbers out loud.

The same was done by Mrs. Kulagina's spouse, who measured her pulse. At the initial moment my pulse was in the vicinity of 70 beats per minute; hers was 80 beats. A minute later my pulse reached 90 beats. Kulagina's reached 110. When my pulse rose to 120 (hers at that time was 150), I stopped the experiment. Mrs. Kulagina suggested repeating it with a device that she had in the house. This time Misha became the experimentee. We joined his hand with this device, which could transmit a signal to a recorder, recording the signal on the tape. The previous result was repeated. For a long time I kept that tape, the matchbox, and the paper with a sketch of its positions. We agreed with Mrs. Kulagina about future experiments and went away with confidence in the reliability of the results. Misha, the former "Doubting Thomas," became a passionate fan of Mrs. Kulagina, and I started to prepare further experiments. However, when I called Mrs. Kulagina two weeks later to arrange the next meeting, she apologized and said that Mr. Sergeev (the associate professor), who paid her from some sources, had forbidden our meeting, and she could not disobey.

At that, my direct communication with Nina Sergeevna Kulagina stopped, but it proceeded indirectly. From time to time I scripted educational movies on physics for the Leningrad Studio of Scientific and Popular Movies (Lennauchfilm), and was also invited there as a scientific consultant.

A few months after my experiments with Mrs. Kulagina, I learned that Lennauchfilm had produced a movie about Kulagina. I knew the operators and the film director, and I was invited to preview it. The movie was very informative. Its creators, as I learned from them, had not really believed in the existence of the paranormal phenomena at first, but during the process of filming they had become convinced of its reality and had done everything possible to bring this belief to the cinema audience.

Mrs. Kulagina's spouse, having brought her to the shootings, had not been allowed into the filmmaking pavilion. Shootings were conducted by several cameras at once and from different vantage points.During an experiment in which Mrs. Kulagina raised a ping-pong ball from the floor without touching it, one of the operators passed his hand several times through the air between the ball and Kulagina's palm while another operator shot this episode. The same was done when she balanced the balance scales, although one of the bowls had a weight of several grams.

There were filmed episodes where Mrs. Kulagina guessed what Leningrad monuments were captured in photographs that were attached to cardboard sheets and held behind her back, with the actual photograph facing away from Mrs. Kulagina.

In this movie Mrs. Kulagina also showed her ability for dowsing. In an area of Leningrad that had been destroyed during World War II and was not yet restored, she found underground communications by means of a wire frame which started rotating in her hand when it was over some of these communications.

For an unknown reason the movie didn't come out and, moreover, was enigmatically withdrawn from the film studio store rooms. It was said that the KGB became interested in Mrs. Kulagina and all research involving her began to be conducted in a certain establishment closed to the general public.

However, by the time Mrs. Kulagina passed away it was possible to find a description of many of the experiments conducted with her on the Internet.

For me, my acquaintance with Mrs. Nina Sergeevna Kulagina was a milestone which has significantly affected my understanding of the regularities of Nature. I believed in the reality of the parapsychological phenomena, and my education and related outlook excluded any supernatural origin of these phenomena and adjusted me toward their natural-science justification.

I continued to work in the field of knowledge given to me by my fate, solved a lot of problems available there, wrote two scientific books, and defended a PhD and doctor's dissertations. But all this time, I always remembered the numerous "why"s that accumulated in my mind. Their lack of resolution incessantly smoldered in my consciousness. I periodically came back to them, but their explanation escaped me each time.

Number 4 - Vadim Polyakov: PhD, Engineer-Scientist Supporting the Health of Astronauts on the Earth Orbit

"Anyone can become extrasense."
Polyakov V. B.

I recall that Professor L.L. Vasiliev, who worked in the field of parapsychology in the 1950s, organized a special laboratory at the Leningrad State University for his research. This lab was located in a former rector's wing, where a famouse Russian poet, Alexander Blok lived at the time with his parents. In one of the rooms of the laboratory floor, the walls and ceiling were completely covered with metal plates to prevent the penetration of electromagnetic radiation. In particular, Mrs. Kulagina was in this room when Professor Vasiliev carried out telepathic experiments with her. I do not know how it is now, but from 1980-1988 this room, as well as Prof. Vasiliev's whole former laboratory, was rented by the Leningrad Polytechnical Institute for the experiments of its Associate Professor Vadim Borisovich Polyakov.

Figure 21. Vadim Polyakov.

Mr. Polyakov was a graduate of the Leningrad Mechanical Institute, had a PhD in Technical Science and headed the Laboratory of Biodynamics at the Leningrad Polytechnic Institute. Mr. Polyakov is the author of the book "Practical psychic". He possessed some psychic abilities, and engaged in treatment and diagnostics. He conducted parapsychological research under a contract with the Flight Research Institute located close to Moscow. His main task was "a contactless maintaining" of Soviet cosmonauts on the orbital space station. When the devices located at the station for monitoring the status of astronauts reported to the Mission Control Center any deviations from the normal pulse rate, blood pressure and other parameters related to the health of astronauts, or they complained of malaise, the Center called Mr. Polyakov and asked him to "support" the astronauts. Mr. Polyakov locked himself in his office and "supported" the astronauts using contactless therapy. 20-30 minutes later a call came from the Center again, this time with one phrase: "Thank you, everything is now o.k."

In one of the experiments carried out by Mr. Polyakov, the collective contactless impact of a group of people on the functional condition of a person was studied. A student volunteer participating in the experiment was in the above-described isolated room, shielded from the penetration of electromagnetic radiation.

Figure 22. The Rector's Wing of the Leningrad State University, located to the left of the gate. There are departments of theoretical physics, geophysics, and the former laboratory of Professor L. L. Vasiliev, in which V. B. Polyakov conducted the experiments described in this book.

Mr. Polyakov, meanwhile, was located in an adjoining room with a group of observers, and participants of the experiment (of which I was one). All of them were asked speculatively (mentally) for a fixed time to send signals designed to increase the student's heartbeat and to raise the temperature in certain parts of his body. The student was covered by sensors, and devices wrote down any fluctuations of pulse and temperature of parts of his body over time. Some notable deviations from a standard state (i.e., when signals were not sent) were recorded in these experiments.

In another experiment, Mr. Polyakov demonstrated (on participants) his ability to make a contactless tangible impact on a human. He stood about a meter away from me and his index fingernail scratched my left shoulder, without touching it. His finger was approximately 30 cm away from my shoulder. Then he asked each of us to do the same to ourselves, imagining that the finger – although not touching a shoulder – was extended and left a scratch in the corresponding place. It seemed to me that I felt the touch of my finger, but the feeling was not very convincing. At the same time, some of the other participants of this experiment enthusiastically declared its success.

Once, I asked permission from Mr. Polyakov to bring one of my friends to him for a treatment session; he was suffering from sciatica at that time. The friend was a professor of physics, Budtov Vladlen Petrovich, skeptical about the real existence of paranormal phenomena. However, his painful condition forced him to agree to go with me. I wanted to help him get rid of the pain that was torturing him and to ensure that the method of treatment, which Mr. Polyakov (in his stories) owns, was not a figment of his imagination.

Dr. Budtov was suffering from the latest in a series of attacks of radiculitis, so strong that he could not completely straighten. After approximately three minutes, during which Polyakov's fingers were clutching one of Budtov's ankles, I noticed that Mr. Budtov turned pale, and I had a feeling that he was barely on his feet. I asked him if this was true and received a confirmation. (All of this time, Mr. Budtov had stood up, because the pain in the lower part of his back did not allow him to sit down). Then Polyakov (who could not see the condition of his patient, as he was at his feet and was focused on them) apologized and said that he had overdone it and had apparently underestimated the potential effect. He suggested that Mr. Budtov should sit on a chair and asked him whether he should continue the treatment, or whether it should be discontinued. In response, Dr. Budtov, wincing in pain, sat on a chair and asked to continue the session. He knew that he was not only trying to get rid of sciatica, but at the same time he was conducting an experiment on himself; he knew that he was running a risk by agreeing, but like a true scientist, he decided to continue the experiment to the end. After a few more minutes, Polyakov completed the treatment and asked Dr. Budtov to get up and straighten himself. The pain had disappeared. The attack had gone and was not repeated. Dr. Budtov was rewarded for his patience and his courage!

Number 5 - Vladimir Shpunt: Physicist, PhD, Alternative Medicine Specialist

In 1989 I moved with my family to America and continued my academic career, first as a professor of Purdue University, and then at various so-called high-tech chemical, biological and pharmaceutical companies, where I was developing physical methods for facilitating the discovery of new drugs. To my initial education in theoretical physics was added understanding of the specifics of flexible-chain macromolecules, proteins, DNA, and numerous entities which play an important role in organic and inorganic life, and in many of the branches related to physics, chemistry, biology, biochemistry, pharmacology and medicine. I felt a growing need to bring together this knowledge, and fate brought me to a unique person, Vladimir Haimovich Shpunt , PhD. He is (as I am) a doctor of physico-mathematical sciences from the Ioffe Physical and Technical Institute of the Academy of Sciences of the USSR in Leningrad. I found in him a kindred spirit, and for 20-plus years I am enjoying his company. In addition to a deep understanding of the physical foundations of the Universe, he has extensive knowledge in biology and eastern medicine and has a rare gift from Nature, the ability to maintain people's health and in some cases even treat various ailments.

When Vladimir puts his hands to your body, you feel them as two flat irons, with a temperature in the range of 38-42 degrees Celsius. At the same time, the temperature of his hands themselves is normal (36.6°C), i.e., the high temperature heat that you feel comes mainly from within your body. Vladimir cannot, for example, increase the temperature of a thermometer.Vladimir's hands not only increase the temperature inside the patient's body. They also have a great healing effect that spreads throughout the whole body of the patient, improving the state of all of his or her internal organs. But it is not only the touching (contacting) with his hands that has a therapeutic effect. Vladimir can also treat a patient without any contact, at a distance from him. The patient may be right next door, or in another city, or on another continent. The effect is the same. To achieve it, it is enough for Vladimir to think about the patient with love, to focus his attention on the patient, to "include" the patient into his "system," and to wish him a recovery. Someone may see this as a prayer. And here, indeed, we have something in common. Later I will return to this issue. And now I will say that I have experienced both of Vladimir's treatments (with contact and without it), both for myself and for the people who are close to me, and I have solid evidence of the effectiveness of both of these treatments.

Figure 23. “The mainstay of treatment is love.” Vladimir Shpunt.

People from different countries have been (and are still) among Shpunt’s patients. These countries range from Russia to Indochina. Among them are the US, the UK, Germany, Finland, Israel, Vietnam, China, and the former republics of the Soviet Union: Russian Federation, Georgia, Armenia, Azerbaijan and the former Soviet republics of Central Asia. These patients were suffering from nervous disorders, eye disease, cancer and many other diseases. Some of them have been completely cured by Mr. Shpunt, others have significantly improved their states and extended their lives.

Vladimir Shpunt is the founder and president of the advisory company “Optimization & Technology Corporation.” He carries out research in the field of physics of the solid state different objects, studies switching and memory effects in disordered materials, including the electrical, optical and morphological properties of human skin (Mr. Shpunt started these investigations [54-56] while still working in the Ioffe Physical and Technical Institute in Leningrad), and provides medical help with the alternative medicine methods that are carried out here. Vladimir simply maintains the vitality of many people and helps them in different situations.

It is important that the perceived effect of the contactless treatment conducted by Dr. Shpunt is the same as the treatment with contact. The body of the patient is warmed and then recovering action takes place. Obviously, there can be no question about the transfer of some energy at an arbitrary distance, by means of a certain field. Vladimir, working with the patient, does not feel any loss of his energy. And the conclusion is that the basis of both treatments, both with patient contact and the contactless one, has the same mechanism.

What kind of mechanism is this? How is information transmitted from the healer to the patient? Why is the same treatment method a universal remedy for various diseases? Why do so many "why"s (rooted in our distant childhood) remain unresolved?

I am now already past 70, and my experience, my knowledge of theoretical physics, and my belief in the reality of paranormal phenomena and the effectiveness of methods of alternative medicine offer hope in a possible merger of such different aspects of life on a single scientific basis. The extraordinary natural gift of Vladimir Haimovich Shpunt serves this hope as an unquenchable guiding star.

Figure 24. Me (on left) and V. Shpunt discussing the results of some experiments.

III. Thermodynamic Approach as a Link between Medicine and Its Alternative Options

1. Placebo Effect from the Point of View of the Thermodynamic Approach

A. The Placebo Effect in Medicine

In medicine, a phenomenon called the placebo effect improves human health due to the fact that the patient believes in the efficacy of some influence, which is in fact neutral. Besides receiving a neutral drug, such an effect can be achieved, for example, by performing certain procedures or exercises, the direct effect of which is not observed.

In other words, the placebo effect is based on auto-suggestion. It has been known since ancient times. However, any explanation of it does not exist. How does a person's belief in the recovering effect of treatment help to cure him? A possible answer can be found in the thermodynamic approach used here. For this, we must once again turn to the equation G = H-sT, where the Gibbs free energy G has a minimum value at the equilibrium of an isolated thermodynamic system. The probability of this state W is determined by the G value according to the following equation

$$W = \exp(-G/kt)$$

It is known that the thinking process consumes a lot of internal energy, H. The energy is spent on certain biochemical processes that accompany the process of thinking. The brain transfers from a calm state of equilibrium to an excited state of non-equilibrium. The brain's entropy, s, is reduced. The perturbation must go away to return the brain to the state of equilibrium.

According to E. Schrödinger [50], the thought is a certain degree of ordering of some material object. In our thermodynamic approach this object is that part of a brain which is responsible for this thought. This means that on the chosen site of a brain, perhaps, a temporary structure is formed by a group of neurons, a difference of corresponding electric potentials and the related biocurrent. As a result of this structuring (or, speaking Schrödinger's language, these orderings), the entropy of this site goes down, as well as the entropy of the entire brain and the whole organism. After realization of the thought, or its cancellation, this temporary structure disappears (seemingly thaws) and the organism comes back to its initial state of quasi-equilibrium.

When a sick person believes that the treatment will help him get well, thinking about this does not leave him even for a minute. The brain is all the time (consciously or unconsciously) in a state of excitement in expectation of the recovery process, as if listening to it. That is, the brain is in a state of non-equilibrium, and can only get out of it after treatment, when this perturbation subsides.

How can the organism return the brain to a state of equilibrium? Only by strengthening the fight against the disease. And this is what it does, eliminating the reduction of entropy in the patient's body. The disease goes away. The brain receives an appropriate signal. The perturbation of the brain disappears. Its entropy increases. The thermodynamic equilibrium is restored. Self-hypnosis has worked. This is the basic principle of the placebo effect.

The drugs used to treat the disease increase the entropy of the patient's body. That is, the cumulative effect of the drugs' action and the placebo effect is the basis of any treatment process.

There is another, less obvious action of the placebo effect on the healing process. For example, let us consider what happens when a person gets a bruise. What is his first reaction? First of all, he instinctively puts his hand to the injured spot. Often, he rubs this place with his hand. This behavior is programmed into human nature. The person unconsciously believes that by putting a hand to the injured spot, he first reduces the pain sensation and, secondly, the "cure" of the injury contributes to the elimination of the bruise's consequences. This is an unconditioned reflex, resulting in this case in the placebo effect. And what is the result? It is small. But still, by rubbing his hand over the bruised spot, the person unconsciously believes that, to some extent, this mitigates the effects of the injury, and he really feels some relief. Why does this happen? First of all, the person additionally focuses on the injured spot, waiting for relief. In the brain there is an additional perturbation, i.e., an additional decrease in entropy. The brain, already released from a state of equilibrium after the injury, is even more removed from this state. At the same time, in the area of injury there is further reduced entropy, as a result of the application of rubbing the bruised place with a hand, as well as of focusing the person's attention on the place. This decrease in entropy does not aggravate an injury, but the body reacts to this decrease in entropy, by increasing activity in the aftermath of injury. As a result of this activity, the entropic component of the Gibbs free energy on the bruised area of the body increases more intensively, the excitement of the brain decreases and the thermodynamic equilibrium is restored faster.

A similar behavior is demonstrated by surgical patients in hospitals. When walking during the post-surgery period, they will automatically, without thinking about it, keep a hand on the operated area of the body and thus, without even knowing it, contribute to a more rapid recovery.

Apparently, a similar method is used by animals for self-treatment. They have no hands, so they use their tongue. When an animal licks a wound, it not only washes it with saliva, but also turns on an additional entropic effect, described above, leading to the more rapid healing of wounds. I have forever remembered a story told by a magnificent surgeon, the friend of my parents, Professor Nadezhda Ivanovna Volodko. During that time I was already studying at the University. Nadezhda Ivanovna came to us in a very excited state, which was very rare for her. "Can you imagine?" she began her story. "Half a year ago, I did a surgery on a patient from a nearby village. He looked a strong man of about 50 years old. The disease symptoms of which he complained, a preliminary examination and a subsequent x-ray analysis indicated impassibility in his gastrointestinal path. The patient was placed in a hospital room and began to be prepared for the surgery. A few days later, he laid under a general anesthesia on my surgical operating table. When I opened his body in the corresponding place, I saw metastasis of cancer cells in such a quantity that if they were cut out, nothing would function there. It was senseless to continue the operation. I sewed him back up. The patient was returned to his hospital room and when the anesthesia was gone, I visited him and, as usual in such cases, told him that the surgery had taken place successfully and that he would soon be sent home. In a week he was discharged from hospital. He felt approximately the same as before the surgery, which in reality had not taken place.

He left hospital in high spirits, but I felt very awkward. Today, this patient, full of vital forces, had been waiting for me since morning at the entrance to the hospital. On his shoulders he had a bag full of gifts which he had brought to me as in gratitude for his recovery. I didn't immediately recognize him, but having recognized him I could hardly stand up, I was so surprised. In my understanding, this person had returned from the afterlife. I quickly took him to my office, examined him, and led him to an X-ray. He was absolutely healthy! The gastrointestinal path was clear! I simply stared in disbelief!"

I think that many of us can remember similar stories about wonderful recoveries from apparently desperate situations. However, this case is a special one. I heard that story not from just anybody, but from the primary source, from a person whom I absolutely trust.

So, the placebo is an effect of auto-suggestion. It is possible to give many examples of when auto-suggestion led to positive results. At the same time, there are many examples of the negative influence of auto-suggestion. If a person looks pessimistically at his treatment, if he doesn't believe in this treatment, it can significantly disturb the medical process and even nullify it. A negative state of mind in the consciousness of the patient, as well as a positive one, has an entropic nature. From the point of view of entropy, it is not important to what side this state of mind is directed. There is always a certain perturbation of the brain, which is cancelled only at the realization of this state of mind. Only by the elimination of the perturbation can the organism return to a state close to an equilibrium. Otherwise, it only moves away from it.

B. Effect of Placebo in Sport

> "Victories and defeats are forged not in muscles, but in brains."
> Bob Beamon, Olympic champion.

The effect of placebo works not only in medicine, but also in many other aspects of life; in particular, in sports where the entropic perturbation, which arises in the brain of an athlete, relates to the functioning of muscles in his body, their force, endurance, and mutual coordination. Here are some examples from various types of sports.

In my first year at university, I went in for gymnastics. During that time teams did not have psychologists; their role was to some extent carried out by coaches. Our coach, a sports master called Gena Mitrofanov, according to his primary education was a medical doctor. He told us that during lectures at his Medical Institute he had often thought over his future performances at competitions. Thus, his attention completely switched over to "scrolling" in his head some combinations on various gymnastics equipment. Sometimes, when the lecturer noticed that the student was not listening to him, he made some remark and, knowing Gena's passion for sports (and Gena was then the champion of the city and a candidate for the national team of the country), the lecturer (with a smile) suggested to Gena "to get down" from a bar, rings or parallel bars and to switch over to medicine.

And so, from Gena's stories we learned that it is necessary to prepare for competitions not only physically, but also psychologically. Moreover, this preparation has to begin long before the competitions (do you remember the episode with the lecture?) and come to an end only at the place of the event, just seconds prior to the performance. The last stage of the preparation looks so:

The gymnast comes to the corresponding gymnastic equipment and for about a minute he or she stands still, with eyes closed. During this time he or she "scrolls" in his or her head through the whole chain of combinations which is to be executed. The chain "is scrolled." In the gymnast's brain a structure appears which will control his or her performance.

By means of this peculiar "self-adjustment of mind" he or she can bring himself or herself out of a state of thermodynamic equilibrium. He or she can return to it again, only having executed the chain of combinations conceived and imprinted in his or her head.

The confidence in his or her forces and the self-adjustment of mind often help the athlete to achieve outstanding results. For example, in speed skating there was always a classification of sprinters and stayers. Sprinters run slower than stayers on long distances (5000м and 10000м), and stayers run slower than sprinters on short distances (500m and 1500m). But at one time, Eric Heiden (an American athlete) made a switch into speed skating from cycling. Cyclists are not classified as sprinters and stayers. They have to have both speed and stamina. Eric was sure that this was also possible in speed skating. He considered that if someone could show a record result, it would mean that this was not outside of human abilities and that with the appropriate preparation any athlete with the corresponding physical abilities could repeat this result and even exceed it, at least a little. And Eric proved it, by winning at the Winter Olympic Games in Lake Placid in 1980 all distances: 500, 1000, 1500, 5000, and 10000 meters, and having set 4 Olympic records and one World record. Only thus could Eric Heiden remove the entropy disturbance of his brain and return to the state of equilibrium he had enjoyed preceding his transition to speed skating. Eric Heiden's achievement was – and after more than 30 years, remains – unprecedented in the history of speed skating.

Persistent training allows an athlete not only to increase his skill, to build muscles and so forth, but also to strengthen a belief in his own capabilities. The thought that someone can lift a record weight, and that he is already close to lifting this weight, and is now properly trained, allows the athlete to suggest himself that he (or she) can cope with this weight and even exceed it a little. And this auto-suggestion, leading the athlete from a condition of thermodynamic equilibrium, often works. The athlete manages to slightly raise a record ceiling, i.e., to lift a weight about a kilogram heavier than the world record, to jump one centimeter higher, to run a hundred-meter race one fraction of a second faster, etc. The establishment of the record removes the stress in the brain of the athlete, increases its entropy and returns it to a state of equilibrium.

The schedule of the establishment of World records in different types of sports visually confirms the correlation of auto-suggestion of athletes and the results they have reached. We will track it using the example of weightlifting. From 1970 to 1977 Vasily Alekseev was a super-heavyweight. His first World record in a push by two hands was reached in January, 1970. Then Alekseev managed to lift a bar weighing 221.5 kg. Over the next 7 years he raised the ceiling of the World record 24 times and brought it up to 236 kg. From competition to competition, the weight of his bar rose by 0.5 kg, 1кг, 1.5 kg, 2 kg, etc. All this time, Alekseev persistently trained and strengthened in himself the confidence that at the next competition he would be able to lift a weight that would be at least slightly heavier. This auto-suggestion worked!

Here is another example from female track and field athletics. A pole vaulter, Yelena Isinbayeva, broke World records 28 times during the period from 2003 to 2009, having raised the bar from 4.82 m to 5.06 m. Each time she managed to improve the record result by 1, 2, or 3 cm. As well as Alekseev, she believed that, having properly trained, she would be able to overcome the level at a height slightly exceeding her former achievement – and she managed to do it!

The example of Isinbayeva is also remarkable in that her physical data looked insufficient for a pole vaulter. See yourself: her height is only 174 cm, and her weight is 65 kg. At 15 years old, Yelena was removed from the Olympic reserve school, as an unpromising jumper. But she didn't want to leave sports, so she found a coach who agreed to take her on and set herself a goal to achieve significant results. As a result, within several years the entropy of her brain (and consequently of her whole organism) was lowered. Her organism had been brought out of an equilibrium state and (following Le Chatelier's Principle) sought to return to it. For this purpose, Isinbayeva's body (muscles, coordination of movements, skills in using the pole) had to realize the conceived goal, and as we see, she realized it! Auto-suggestion has worked! This is a fine illustration the implementation of the placebo effect in sports!

The last example I would like to give is also from track and field athletics. This is a fantastically long jump made by Bob Beamon. His sports career is, to a certain degree, similar to Isinbayeva's career. Bob Beamon's sports progress was accompanied by vital disorders, dissonance in his family, and money problems. Before the Olympic Games he was excluded from the athletic team of El Paso University and deprived of a grant for participation in a political protest. As a result, Beamon` remained without a coach. In the Olympic team of the USA he was the second after Ralf Boston. No great expectations were placed upon him. However, aside from his great passion for it, sports for Beamon, as well as for many other black athletes, represented his only chance to be a respectable person. Beamon participated in commercial competitions, which did not bring him a big income. His financial difficulties were aggravated by an unsuccessful marriage, but he trusted his star and adjusted himself for victory. The competitions in Mexico City began unsuccessfully for Beamon. At the qualification standard of 7.65m he made mistakes in the first two qualification attempts. However, at the third attempt (taking the advice of Ralf Boston) he "secured himself," and reached the final with the second result, 8.19 m. Ralf Boston, who had achieved 8.27 m, was the best in the qualification.

On the day of the finals it was drizzling. Beamon was the fourth in the list. On the first attempt, his running start turned out extremely fast, exceeding the ideal level. The jump was very high (more than 1.5 m at the highest point), and his landing was powerful and rigid. Beamon pushed himself away from the sand with both feet and jumped out of the hole.

In his own words, he felt that the jump had turned out successful, and he stayed in a good mood. It was not in his rules to watch the results of his rivals, or his own results. "Victories and defeats are forged not in muscles, but in brains. Brain is a huge factor in the Olympic competitions. If I had gone to see the other jumps, for me it would mean that I am not self-assured. And uncertainty brings defeats."

Having read these lines, someone will probably call the entropy idea, stated here, a plagiarism (only with the use of different terminology), so similar is Beamon's statement to this concept. Yes, Beamon's brain had been brought out of a state of equilibrium by an immense desire to achieve great publicity in sports. His entire organism was adjusted to this success. And the success came! Thermodynamic balance was restored!

Examples of such results being reached without any training, and even by non-athletes, are known. In an excited state, when there is no time for thinking, and it is necessary to act immediately – for example, running away from a pursuer – a person can show remarkable force; he can jump over a very tall fence or the widest crevice. In this case, everything is made, as they say, in one breath. Without understanding why, this person takes the necessary action, cancels a sharp decrease in entropy, connected with the danger threatening him, spends the maximum quantity of energy possible, reaches the desired effect and, as a result, to some extent, restores the thermodynamic balance previously lost.'

2. Alternative Medicine from the Point of View of the Thermodynamic Approach

Every year the contribution to human health made by alternative (or unconventional) medicine increases. This is the collective name for methods with abilities to prevent and to treat various diseases without using medical drugs. Healing, homeopathy, acupuncture, treatment by magnetic fields, hypnosis, chiropractic, yoga, reiki, mustard plasters and many other things belong to this category. In this section I will try to describe some of these methods and give a possible explanation for their medical influences.

A. Manual (Contact) Therapy

"A healthy body is defined by a healthy mind"
Mikao Usui, founder of Reiki system

It is possible to explain by means of entropic interaction the treatment methods of manual therapy, i.e., the use of the hands for the diagnostics and treatment of structural and functional deviations in various fabrics and body organs. Starting the treatment, the person possessing an ability to carry out such treatment (he or she is usually called a healer) focuses the attention on the patient's organism. Thereby, the healer and the patient are united into one thermodynamic system. This system automatically becomes non-uniform according to many characteristics, including also the distribution in it of entropy, as the value of the healer's entropy is different from that of the patient. The entropy heterogeneity instantly leads to a decrease in the entropy of the whole system in comparison with the total entropy of the healer and of the patient before their interaction. In response, the organisms of both subjects seek to increase the entropy, in particular, turning on the biochemical processes promoting the recovery of the patient.

At the same time, one more effect starts working. When the healer puts his hands onto the body of the patient, the latter immediately feels heat proceeding from the healer's hands. At the same time, the temperature of the hands does not differ from normal (about 37°C). However, the hands of the healer strain, i.e., they are structured. They radiate excessive heat (like liquid, when freezing), without changing their temperature. The structure (caused by a disease) receiving this heat "thaws" and disappears. The corresponding measurements of this heat shows that it is radiated in the narrow range of infrared radiation (around 10 microns). Probably, this range has the best medical effect, "melting" the structural formations related to the disease. Furthermore, under the hands of the healer, in the body of the patient there is an additional fall of entropy, connected with a specific structuring (caused by the placing of the hands), which also radiates excessive heat. This is very similar to treatment with the help of "mustard plasters." This local reduction of entropy is classified by the organism of the patient as an undesirable development of the disease, thus stimulating a more intensive fight against it, waged by the organism, which, trying to restore the broken thermodynamic balance, "turns on" new biochemical processes (and intensifies those which are, probably, already taking place), which leads, in particular, to a temperature increase and, as a possible consequence of this, to the elimination of pathogenic structural formations. In other words, in treatment by means of imposing a healer's hands it is possible to point to four complementary effects:

1. Placebo effect connected with the belief of the patient in treatment by a healer.
2. Heat coming from the hands of the healer "thaws" the structure caused by a disease.
3. The fall of entropy connected with the imposing of hands leads to a response by the organism of the patient. The response is expressed in the strengthening of the activity of its immune system. Blood inflow amplifies to the place of the imposing of hands. Biochemical reactions, by means of which the organism treats the disease, are intensified. Probably, additional reactions appear, in particular those which bring heat to the structure caused by a disease. Under the influence of this heat, the structure "thaws," as well as from heat received from the hands of the healer. The intensification of other medical actions is not excluded in the organism of the patient.
4. The entropy decrease is connected with its non-uniform distribution in the healer-patient system.

A correct understanding of the entropy effects occurring in the body of a person can be used for the treatment of various diseases and injuries. Adding the corresponding heat and additional artificial fall of entropy (or of an entropy component of the free energy) on a sore area of the body will bring, on the one hand, to the "fusion" of undesirable structural formations, and on the other hand, it will cause a corresponding reaction of the organism which, as a result, will more intensively "treat" the fabrics of this site until their absolute recovery is achieved.

It is not excluded that for such a treatment it is possible to use not only the hands of people possessing the ability to treat by radiating corresponding heat. If we could replace them with an electromagnetic oscillator, radiating infrared radiation in the range close to 10 microns, it would probably be possible, at least partially, to achieve the same curative effect. The thermal influence that we experience when visiting saunas and steam rooms undoubtedly includes an infrared range with a wave length of 10 microns and therefore makes a refreshing and salutary impact.

B. Treatment by Structured Water. Tatiana Trofimowa (No. 6 of "The Magnificent Seven")

Once, being at a scientific conference in San Francisco, I incidentally learned about a surprising healer living nearby in Silicon Valley. It was Tatiana Trofimowa, whose vital motto is used as the epigraph of this section. She has a medical education, healing abilities, a striking imagination, and an exclusively kind heart. All day long Tatiana accepts patients with various diseases, successfully healing them using methods of contact and contactless therapy. Her contactless diagnostics of diseases is phenomenal! I say this because I was influenced by it and then tested it on my relatives.

Figure 25. "To live, without being ill!" Tatiana Trofimowa.

Then, at the end of the conference, my wife Anna and I went to meet Tatiana, to get acquainted with her, to ask her about the technique she used and possibly to test it. By this time, I was already aware of trying not to miss any opportunity to find any information available about so-called alternative medicine.

Tatiana turned out to be a friendly woman with a pleasant exterior and a very strong character. Having looked at me for several seconds, she said that she did not see in me any serious problems with my health. However, she said that my heart was increased in size and its left ventricle was larger than the right one. Furthermore, she noted that something was wrong with my left knee.

I was literally shaken. Tatiana's diagnostics was very precise. But to receive such information about the condition of my heart, my doctors had to conduct rather difficult investigations, which they had done some time ago. And in general, nobody paid any attention to any problems with my knee. Without complaining, I annually participated in mountain walks, skied, and played soccer. In actual fact, the knee had once been really injured. It occurred when I was about 25, during one of my football matches. But over time I forgot about this trauma, as it did not affect me in my current life.

After that, Tatiana greeted Anna and also astounded her with the diagnostics. It so happened that there were no other patients for some time, so we could communicate with Tatiana in more detail. It turned out that the touch of her hands causes a thermal effect, comparable to that of our friend, Vladimir Shpunt. Tatiana said that she treats many patients who are far away from her office, and showed us an album with their photos, on which she focuses her attention during sessions of such contactless treatment.

Tatiana's treatment with the help of water should also be considered contactless. Tatiana structures the water, focusing her attention on it. In my understanding, the entropy perturbation created by Tatiana inside herself, is transferred to the drinking water which, under Tatiana's influence, is included together with her in one thermodynamic system. The water can thus be in any suitable container. It is structured, in the sense that its entropy goes down. Tatiana is kind of programming this water, wishing that this structure of the water should remain for a long time. When the patient drinks the structured water, it gets mixed with the water which is already in the body. As a result, the structured water penetrates into all parts of the body (both sick and healthy), reducing their entropy. The body of the patient becomes even more out of balance (it has already gone out of balance as a result of the disease), and the body begins to heal itself more intensively, increasing the entropy and restoring a thermodynamic equilibrium. They say that nobody is absolutely healthy. Therefore, this structured water is beneficial to everybody. It has a healing effect, refreshing and rejuvenating. Compresses with structured water promote the faster healing of wounds and injuries.

The effect of the structured water is not merely speculative. It can easily be seen by comparing normal drinking water with the water structured by Tatiana's hands.

Figures 26 and 27 show photos of water, frozen at -80°C. The photos were obtained using a microscope. Water from the same bottle was poured into two arbitrary vessels. One of these vessels with water was exposed to the action of Tatiana's hand. The other vessel with water was not exposed. Then both vessels were placed into a refrigerator with the already mentioned temperature of -80° C. After 30 min the vessels were removed from the refrigerator, placed under a microscope and photographed. Air bubbles in these two vessels with frozen water are clearly visible on the pictures. In the container not exposed to Tatiana's hands, these bubbles are distributed randomly, and their shape is nearly spherical. But in the container with water touched by Tatiana's hands, the bubbles are elongated and oriented in one certain direction, i.e., this water is polarized, or, as we have seen, structured, and the entropy of water in the vessel is reduced compared to that in the other vessel.

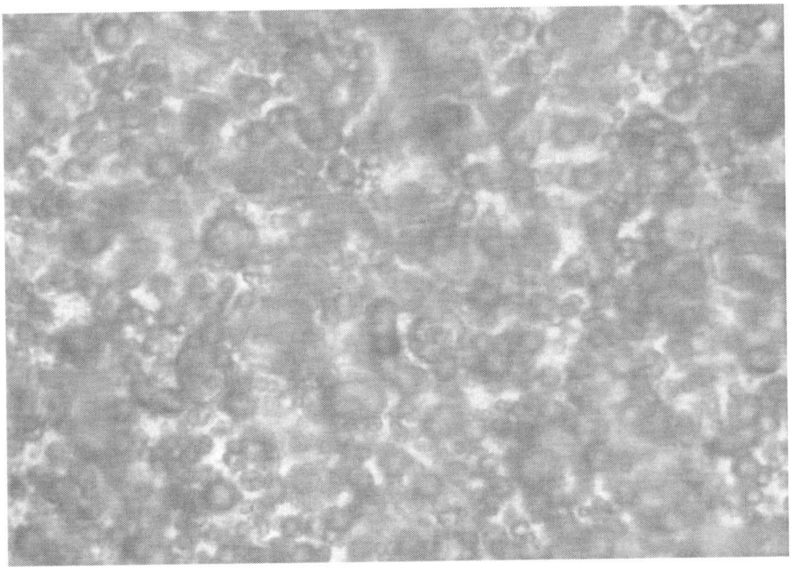

Figure 26. Picture of frozen water in the vessel that was not touched by Tatiana.

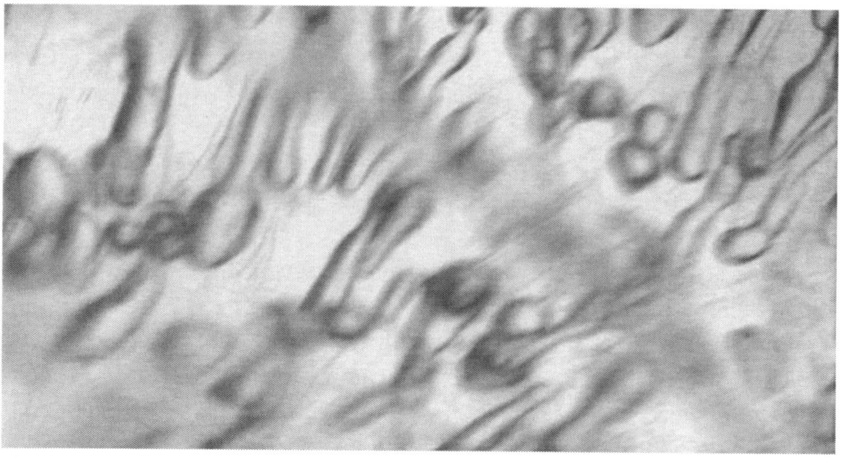

Figure 27. Picture of frozen water in the vessel that was exposed Tatianato Tatiana's touch.

The procedures described here, including photography of the frozen water, were carried out repeatedly by the author personally in the biological laboratory of the Brandeis University and repeated in the laboratory of the pharmaceutical company Hoffman La Roche. I checked the action of compresses with the structured water on myself. I chose compresses because their action can be checked more convincingly. With their help, within several days I managed to get rid of blefaritis of the eyes, which ophthalmologists had been unable to cure for several years. It was also successful in cleaning pigmentary age spots from my face. These compresses effectively help at the emergence of catarrhal herpes, and significantly reduce the period needed for healing wounds, scratches, bruises, etc.

The technique of the artificial reduction of entropy has long been used in sports medicine for the treatment of injuries. According to this method, the injured muscle is first cooled and then the cooling is replaced by a hot compress. Upon cooling, the entropy component of free energy, sT, is lowered, bringing the injured muscle even more out of balance. The body resists it. Since the applied external cooling is not controlled by the body, the latter tries to prevent further decrease in entropy, in part by slowing down the process of tumor development associated with the trauma. When this development stops, the cooling must be replaced by heating. At this stage, the entropic free energy component, sT, increases at the expense of increasing the temperature T. At the same time, the structuring caused by the injury "melts" or disappears. The muscle is returned to the state of thermodynamic equilibrium (the initial value of the free energy G is restored) and the healing process of the injury is completed.

C. Homeopathy

Homeopathy is a type of alternative medicine which is based on the principle of “treating a similar object with something similar.” This principle suggests the use of highly dissolved drugs that cause symptoms in healthy individuals similar to the symptoms of the unhealthy patient. On the one hand, the field of official medicine considers this principle scientifically unproven, but on the other, it uses the analogy of that principle by making vaccinations against infectious diseases by producing vaccines of attenuated or killed microorganisms, of their metabolic products, or of their antigens produced with the help of genetic engineering or chemical means. The “entropy approach” explains the therapeutic effect of homeopathic medicines by the additional decrease in the entropy of the diseased organ of the patient. Thus, there is a stimulation of the immune system, which, seeking to increase the entropy, begins to fight the disease more actively.

D. Acupuncture

Medical treatment measures work variously, but many of them use entropic interactions. Their effect is summarized with the “entropy correction” from the organism. Acupuncture can serve as one such example. The action of this method of treatment is defined by the existence on the human body of special points which are connected with various organs: heart, liver, lungs, stomach, etc. Modern medicine knows about two thousand such points. They are called biologically active points. By influencing these points, it is possible to treat diseases, without applying any drugs. For this purpose it is necessary to simply press the corresponding point, to massage it, or to introduce a needle into it. When the needle is introduced into a certain point (of a human body), connected with the sick organ, there is a local reduction of entropy (probably, very sharp). The organism reacts to it as to strengthening of the disease of the appropriate body organ and starts spending (for the fight against this disease) additional forces; in particular, it increases blood inflow to this part of the body. Actually, the level of the disease does not change because of the needle introduction, but the organism starts treating it more intensively, so the process of recovery can be significantly accelerated.

E. Mustard Plasters

Mustard plasters, which successfully help in the fight against the cold, can serve as another example of the "entropy treatment." How do they work? The interaction of the mustard plasters with the body leads to the local reduction of entropy, s. A certain structure appears in the area where the mustard plaster is placed on the body. The entropy component (-sT) of the free energy decreases, and the value of the free energy increases. The organism qualifies it as a strengthening of the disease and starts fighting against it. It initiates biochemical reactions bringing (beside other things) a temperature increase in the area of the mustard plaster and an additional inflow of blood to that place. These places generate excessive heat, similar to the case of structuring liquids at their freezing. This heat is used for the elimination of the undesirable structure (caused by the disease), as though "thawing". The entropy component of free energy increases and thus there is the desirable medical action.

Here it is necessary to emphasize that mustard plasters do not transfer heat to the body. It appears (it is generated) locally in the body at the expense of its internal resources, as a result of its aspiration to restore the thermodynamic balance (to leave the value of G unchanged).

F. Cupping Therapy

Cupping therapy works using approximately the same principle as mustard plasters. This is an ancient Chinese form of alternative medicine in which a local suction is created by low pressure that is created using heat (fire) or mechanical devices (hand or electrical pumps). This mobilizes the suction of blood into the body tissue. The inflow of blood is accompanied by a local temperature increase and entropy reduction. The effect of it is the intensive "self-treatment" of the organism.

G. Palming

Some known medical methods try to use the salutary influence of hands. For example, about 100 years ago an American ophthalmologist, William Horatio Bates, offered a non-drug method for the improvement of sight, based on this method. He called it "palming," proceeding from the English word "a palm." Bates described the method in the book "The Cure of Imperfect Eyesight by Treatment Without Glasses," published in 1920. According to Bates, the regular placement of hands on the eyes for 5-10 minutes, accompanied by a number of exercises for eye muscles, can significantly help with far-sightedness, astigmatism, presbyopia (i.e., an anomaly of the refraction of an eye, developing with age, when a person cannot see small print or small objects at close distance), etc.

Irrespective of Bates's motivation, the exercises recommended by him are certainly useful for eye muscles. First, they keep these muscles toned, preventing them from weakening and from not carrying out their functions (as often happens with age). And secondly, thanks to the work of these muscles, blood circulation around the eyes is intensified, which is certainly useful for their functioning.

As for "palming," I can recommend (from my own experience) not simply imposing the hands, but also mentally imagining that certain radiation proceeds from the palms. Imagining this, we focus our attention both on the palms, and on the areas on which they are placed, and thus we promote entropic interaction between them.

I suppose that this leads to the more effective (both local, and general) fall of entropy; the action of the immune system becomes more active, stronger, and the revitalizing effect becomes more notable.

It is necessary to adhere to such execution of the palming procedure in all cases of its application.

H. Reiki

One kind of alternative medicine in which the technique of palming is applied is "Reiki." Knowledge and traditions of Eastern medicine are its cornerstones. Reiki is used for the purpose of diagnostics, treatment and prevention of various diseases, and also for the maintenance of a vigorous spiritual mood and the healthy physical condition of the body.The practice of Reiki was founded in 1922 by a Japanese Buddhist, Mikao Usui. The word "Reiki" consists of two hieroglyphs: "rei" and "ki." In Japanese "rei" means the Universe, spirit, soul of matter. "Ki" means energy, mind, mood. Mr. Usui spent a long time searching for and researching various ways of natural healing and, finally, he founded a school at which he started training people in the technique he had developed. Then the classes of "Reiki" extended worldwide. It is claimed that teachers of "Reiki" can transfer the curing energy "ki" to their patients by means of their hands. The situation of the rooms in which "Reiki" classes are given, the lighting of the classes, the music and so forth are carefully planned so as to create a meditative atmosphere. During the session, the Master of "Reiki" imposes his palms on the body of the patient in so-called "Reiki positions." The Masters assure that the curing energy can be available to everyone by means of the "tune-up" which is carried out by the Master of Reiki.

There is a tendency to try to use "Reiki" for fighting against stresses, fears, emotional frustration, chronic pains and for health improvement as a whole. During a session clients feel a deep relaxation, the warming up of their body, and energy inflow. The salutary action of "palming" and "Reiki" can be explained by the thermodynamic approach described here. Just as treatment by means of the healer, who uses manual contact therapy, there are three supplementary effects working together, one is the decrease of entropy in the sick part of the body and entire organism of the patient, the placebo effect; the radiation from the hands of the healer, and the intensification of the immune system.

I. Contactless Therapy

My acquaintance and subsequent friendship with Dr. Vladimir Shpunt became one of the incentives which induced me to develop the thermodynamic approach to various natural phenomena. From my own experience as well as the examples of my loved ones I was convinced that Vladimir is able to treat people with identical success, whether in direct contact with the patients, or at any distance from them, however far. In both cases the body of the patient is warmed soon after the beginning of a treatment session, and the symptoms of the disease either disappear, or decrease (depending on the character of the disease). For example, a headache passes at once, a cold is dispelled, and serious diseases are gradually cured.

It is difficult to suggest that at contactless therapy any energy is transferred from the healer to the patient. A connection between them is established, presumably, instantly (irrespective of the distance dividing them). The healer feels no energy loss. To assume some presence of fields unknown to science (we will say, biological fields) which we cannot find by means of devices is unreasonable, from the point of view of physics. At the same time, the consideration of the healer-patient, as a quantum entangled system of two objects which form one thermodynamic system, looks tempting.

In this case information on a change in the condition of one of the parts of this system becomes instantly known to the other part of the system and leads to the corresponding change of its state. Entropy naturally applies for a role of the carrier of such information in thermodynamics. When the healer starts treatment, he strains, starts thinking about the patient, and seemingly unites with him (or her) within one thermodynamic system. The entropy of the healer goes down and at the same time the entropy of the whole system, healer-patient, goes down, i.e., there is an entropic interaction between the healer and the patient. The organism of the patient leaves its previously steady state. The immune system of the patient considers the fall in entropy as a strengthening of the disease, and seeks to treat it more actively than it had done prior to the beginning of the medical session, i.e., according to Le Chatelier's principle, biochemical processes start in an organism of the patient, seeking to return it to a healthier state. The organism spends a part of the internal energy (enthalpy), increases the entropy, increases temperature, lowers free energy (following the principle of the least action) and, eventually, improves the state.

IV. Entropic Signaling System in the Biological World

Each biological object as a whole acts as one thermodynamic system. If on one of its parts there is an entropy perturbation, i.e., the entropy of this part changes, then the entropy of the whole object changes at the same time. It is brought out of a state of thermodynamic balance, feels this perturbation and tries to neutralize it, i.e., in this consideration we can speak about an entropic signal system.

According to Le Chatelier's Principle, any physical object opposes a change of state connected with the change of entropy. It tries to find any opportunity to restore its status quo and when it is found, tries to realize it. For this purpose certain organs of biological objects have to receive the corresponding signal and execute it. For a human body the most economical way of transferring such a signal is the use of the central nervous system.

For example, let us imagine that you want to raise your hand. According to this desire, you have a perturbation in your brain. The brain is brought out of a state of equilibrium. At the same time your whole organism is also brought out of a state of equilibrium. To return to this condition, your desire has to be either executed, or cancelled. For the performance of the desire, the muscles of the hand have to receive the relevant information, i.e., they have to know what action is required. The organism transfers this information, sending a signal from the brain to these muscles, using the central nervous system. The role of the signal is carried out by the electric current arising at the expense of the difference in electrical potentials between the corresponding site of the brain and the muscles of the hand. Information is transferred. The hand muscles react to it. The hand is raised. The disturbance in the brain disappears. The thermodynamic balance is restored. And what would occur if the central nervous system had failed, or if it did not exist at all? Then the organism would have had to find another way of transferring the information.

At the origin of life on Earth, according to the current evolutionary scientific concepts, one-cellular organisms first appeared, then two-, three-cellular, and finally, multicellular organisms, which initially did not have a nervous system. The question arises: how did these organisms transmit information from one cell to another? For a multicellular organism to be "informed" about the state of each of its cells, inside the organism there had to be a mechanism for the rapid transmission of information, especially related to, for example, danger. Without this, it is difficult to imagine the survival of an organism in an environment where any part of it could be destroyed by an external force, i.e., there had to be a certain initial alarm system (entropic system) preceding the present central nervous system and alternative to it. Such a system could be based on the entropic information. An entropy change in any site of an organism (where it was harmed) would at the same time cause a change in the entropy of the whole organism, i.e., the organism instantly received a signal of danger and could react to it as appropriate.

This system preceded the present central nervous system and was alternative to it. The peculiar slot-hole structure formed by the closely located cells of the organism could carry out its role. Entropic interaction between various sites of such an organism was implemented by the emergence of a difference in the electric potentials between them, generating electric current which, thanks to the tunnel effect, transferred the necessary information.

Over time, with the development of the central nervous system, such information transfer became less demanded. The use of nervous fibers for this purpose (giving more detailed information) became more preferable.

However, it is reasonable to expect that the alternative alarm system did not disappear completely, and is in a condition of "hibernation." Therefore, if the central nervous system fails, one could try to replace it with an alternative alarm system, having found the way to "waken" it. It could be especially important at such diseases as multiple sclerosis and amyotrophic lateral sclerosis (abbreviated to ALS), at which the signal from the central nervous system (i.e., the brain) does not reach its destination.

How can we re-enable the entropic signaling system? It is possible to offer the following response. The patient, whose central nervous system is faltering in the management of some muscles, has to exercise regularly and repeatedly, trying to make the disobedient muscles fulfill his or her orders. Initially, for a long time, these efforts may fail. And this is natural, since the body will try to pass those orders using the nervous signaling system, which is damaged.

However, a new structure ("structure of the order") will appear at one of the parts of the brain after multiple repeatedly unsuccessful attempts. Respectively, the entropy will decrease there, and the body will go out of equilibrium. To return to this state, a muscle has to obtain a signal, transmitting this order, and then to execute it, thereby increasing the entropy of the organism and returning it to the condition it was in before the disturbance. It is demanded by the principles of Le Chatelier and of least action. The belief of the patient in the ultimate success is the essential condition for turning on the second signaling system. This faith must be cultivated by the patients themselves. In what way? First of all, by coordinating in their minds (in their world view) the scientific approach expounded here with their own understanding of the laws of Nature. Then, by starting to train themselves in the performance of the simplest of orders. When these actions start to be (more or less) fulfilled, it would mean that the second signaling system has started to work. In this case, the belief in the final cure will be strengthened and developed. The second signaling system will increasingly be used by the body to transfer the selected information within the body, and, ultimately, will replace the damaged nervous signal system. It is impossible to say that the disease is cured. But it will be neutralized!

V. Explanation of the Paranormal Phenomena

"The process of scientific discovery -
it is, in effect, a continuous flight from miracles. "
Albert Einstein

1. Hypnosis

The objects of inanimate Nature possess a large number of degrees of freedom at a microscopic level, but when they are considered macroscopically, we see that they are components of one thermodynamic system if they can influence each other, i.e., interact entropically or by means of any physical fields or processes. In living Nature (which we associate with the organic world) there is an additional opportunity. Here, to get into one thermodynamic system, it is enough for two individuals only to think about each other. Moreover, it is enough even for one of them to do it. The successful functioning of contactless therapy convincingly proves it, as an experimental verification of this provision.

An example of the entropic influence of one person on another is an order sent in some form. The person giving the order and the person receiving it are parts of one thermodynamic system. From his earliest years, a child gets used to obeying his parents, then his teachers, and becoming an adult, his seniors in rank. It is as though the person is programmed toward the implicit implementation of these orders although, according to Le Chatelier's Principle, he (or she) would like to remain in his former state and to disobey the order. The entropic perturbation in the brain of the person giving the order is, at the same time, a perturbation of the whole system: "commander-subordinate." After implementation of the order, this perturbation is gone and the balance of the system is restored.

The entropic interaction in this example can seem trivial. It is possible to say that if this order had been executed by a person who is not obliged to submit, then it would impress us and would be uncommon. Well, there are no problems with providing such an example.

A hypnotist possesses an ability to influence the condition of the system: hypnotist-patient. He (or she) creates in his (her) brain an entropic perturbation connected with a desire to realize the orders. The entropy of his (her) brain decreases. In the system consisting of the hypnotist and the hypnotized, the entropy also goes down and the whole system leaves an equilibrium state. The hypnotist turns into the commander, and the person subjected to hypnosis turns into an obedient performer. After the session's termination, the hypnotist - hypnotized system returns to a state of equilibrium.

In the example of hypnosis the effect of suggestion from the hypnotist worked and, to some extent, the effect of auto-suggestion by the hypnotized person (unlike the placebo effect, where only the effect of auto-suggestion worked).

2. Telepathy

Telepathy, i.e., the transfer of thoughts at a distance without the use of any other means of communication, is somewhat similar to hypnosis, but here the order is given not by a voice, but by the mind. As in the case of hypnosis, in a telepathic session both participants should be considered as one thermodynamic system. The brain of the person transferring some information strains. In it, it appears as though there is a certain additional structure. Its entropy decreases. At the same time, the entropy of the whole system, consisting of these two people, also decreases. For this reason, the brain of the person accepting the information feels the perturbation in the brain of the person transferring information. The system is brought out of an equilibrium state. To return to this state, the brain accepting the information considers its various options. According to these, structures appear in the brain (replacing each other) relating to these options. When the correct option has been chosen, the structures connected with the information, in both brains, coincide. The entropy of the system sharply increases. The entropic perturbation goes down and the whole system returns to the state it was in preceding the telepathic session. The result of the session, naturally, depends on the sensitivity (to the described increase of entropy) of the person accepting the information. This sensitivity is not only defined by the specific features of the individual, but also by the state in which he or she is. The ability to abstract himself or herself from the current situation and the related problems is a very essential factor.

Therefore, a reception of telepathic information often happens in a dream or in a trance, when the activity of the brain is reduced and it "is open" for the reception of information. This can explain the interpretation of dreams practiced for centuries, as well as their fulfilment.

3. Clairvoyance

Similarly, clairvoyance, i.e., the ability to see subjects or events beyond the limits available to sight and to hear events outside our hearing, can be explained as an ability to connect to information available to other people. Really, everything that we see and hear appears anyway in our brain, and inherently is its entropic perturbation. The person possessing the ability of clairvoyance, and wishing to see or hear something occurring in a remote place, focuses his or her attention on this place and thus becomes a part of the thermodynamic system including this place and the people who are there. The entropic perturbations taking place in this system directly concern him or her. He or she strains to identify them with certain images and leaves his or her state of equilibrium. Various versions of the events taking place there appear in his or her imagination. When they coincide with reality, tension falls down and a balance is restored. Only the "clairvoyant" (and sometimes a person who is in a condition of special excitement, or trance) can feel the sharp increase in the entropy connected with this coincidence.

It is necessary to also consider as a clairvoyance the mirages which are often observed in deserts, at sea and possibly in cosmic space. Mirages are usually explained as light reflection by a border between air layers, sharply various in density. However, this explanation does not always work. The thermodynamic approach offers an alternative explanation for mirages. The desire to see an oasis in a desert is a very strong entropic perturbation of the brain. It tunes itself up onto the area which is this oasis, but lying outside the horizon. Some people can be in the oasis and the entropic interaction with them allows them to see the oasis with their eyes.

Similarly, it is possible to explain the emergence of "The Flying Dutchman" – the ship seen floating to help people suffering a disaster at sea – as a mirage.

Apparently, the vision (by astronauts) of some "angels" outside the spaceship belongs to the same series of mirages. In 1985, a crew consisting of six people was on board a Soviet space station known as Salue-7. Those were astronauts Leonid Kizim, Oleg Atkov, Vladimir Solovyov, Svetlana Savitskaya, Igor Volk and Vladimir Dzhanibekov. It was the 155th day of flight. The crew was engaged in commonplace tasks, preparing to carry out a series of laboratory experiments. Suddenly, the habitual working rhythm was broken. The station was shrouded in a cloud of strange orange gas. All of the astronauts went completely blind for some time from the bright light. When their sight returned, the crew members distinctly saw seven figures outside one window. The aliens looked like humans, but their sizes were gigantic. They had big wings behind and radiant haloes around their heads. In general, the creatures looked exactly as angels are usually described!

The crew transferred a report on the incident to Earth. Having examined it, officials immediately imposed a signature stamp of "Confidential" on the document (which was finally removed after more than 20 years). All participants of the incident were subjected to various psychological and medical tests which, however, showed that no aberrations were present.

I classify the astronauts' vision of angels as a mirage provoked by the expectation of something unusual in space. Religious dogmas, which are implicitly present in our consciousness, occupy heavens with the corresponding characters who, in this case, looked like angels radiating bright luminescence.

I have not seen mirages in deserts, I have not met with "The Flying Dutchman" on the sea, and have not visited cosmic space. But one autumn evening I happened to see a polar light in Leningrad. The sky over the Peter and Paul Fortress blazed bright colors for a minute or two. It was a gripping show, especially if we consider that a polar light occurs mainly behind the polar circle. As the mass media did not react in any way to this phenomenon, I classified it as my own mirage.

4. Predictions of the Future

From positions of entropic interaction, it is possible to explain future predictions too. There are always experts estimating the possibility of some events, proceeding from their scientific or, simply, logical analysis. The mass media often publish the results of these analyses. These forecasts are realized with some probability. Access to this information (according to the scheme described above) can explain the predictability of these events by a psychic.

The person possessing the ability of clairvoyance, and wishing to predict the future succession of events, focuses his or her attention on them. The consciousness of the people, specializing in the analysis of these events, journalists, radio and television commentators, is also concentrated on them. As a result, the whole group, including our clairvoyant, is included in one thermodynamic system. The entropic perturbations taking place in this system directly concern the clairvoyant as well. He or she strains to identify them with certain images and leaves an equilibrium state. In his or her imagination there are various options of the events and when they coincide with those predicted by experts, the tension falls down and balance is restored. We will emphasize once again that only the "clairvoyant" (and sometimes a person who is in a condition of special excitement, or trance) can feel the sharp increase in the entropy connected with this coincidence.

It is known that Wolf Messing predicted the defeat of Germany in World War II some years prior to its termination. The known American astrologer Jeane L Dixon made the same prediction as well. Recognition of Dixon's unusual abilities allowed her to get a prestigious position as a Special Adviser at the White House. However, many military experts and politicians of that time adhered to just the same point of view, estimating the forces of belligerent parties. Therefore, it is possible to say that such an outcome of the war literally "soared in the air" and, finally, was realized with a certain probability.

In a similar way, it is possible to consider Jeane L Dixon's predictions about John Kennedy's victory in the presidential election of 1960, and about a possible assassination attempt on him in one of the southern states, which was predicted by many political analysts and polls.

5. Telekinesis

A psychic demonstrates the phenomenon of telekinesis. By an effort of will, and some movement of the hands, he or she can move objects, lift them in the air, move cups of lever scales, and rotate the arrow of a magnetic compass. Entropic interaction allows us to explain the observed effects rather simply. The psychic and the object, which he or she influences, make one thermodynamic system in this case. The psychic strains and lowers his or her entropy. The whole psychic-object system is brought out from an equilibrium state. To return to this state, a chain of events, eventually realizing the desire of the psychic, has to happen in the system. For this purpose, for example, a certain part of the electric charges, a variety of which is available in the psychic's body in the form of ions and electrons, has to concentrate in the palms of the psychic. Their electric field produces charges of the opposite sign on the surface of a subject subjected to influence. Under the influence of the corresponding Coulomb force, the subject will start moving in the direction of movement of the palms of the psychic. The magnetic field appearing at the movement of the charged palm will influence the compass arrow, forcing it to rotate together with the palm. As in the considered case, and when the influence of an electric field is excluded (the object is in "the Faraday box), it is possible to offer another option for the realization of the entropic interaction. The molecules of air surrounding the object are, as we know, in random walk, which eventually sets their average distribution in this space and the related density and impulses of the molecules. The probabilistic nature of the random walk assumes the possibility of the violation of this distribution. The probability of this violation is very small but, according to Poincaré's reciprocal theorem, can be calculated. Also, it is possible to assume that this probability is realized in a telekinetic experiment, that the entropic perturbation created by the psychic took place, and the whole object-psychic system could come into a state of greater equilibrium. And this can occur only in two cases: either the psychic will refuse the desire to move the object; or the object will start moving. In the latter case, it is necessary for the molecules of air striking the object from below and from one of the lateral sides to transfer to it a bigger total impulse than the molecules above the object and from its opposite side (i.e., there have to be more molecules incidentally moving up and in the direction of the desirable movement of the object than those moving down and in the opposite direction).

VI. Influence of Entropic Interactions on Events

As explained earlier, according to the Mach Principle, the condition of the Universe as a whole has an impact on the processes happening in each of its separate parts, in every (randomly chosen) spatial system. The opposite is true as well. Changes in each local system influence the condition of the Universe as a whole. Respectively, it is possible to claim that the condition of any subsystem of the given system, in turn, depends on the condition of this system, i.e., any two objects of Nature are mutually connected. They can be considered as one thermodynamic system.

From this position, a person and the world surrounding him (her) represent one peculiar thermodynamic system. The entropic perturbations happening in the world have an essential impact on the condition of the person. They bring him or her out of an equilibrium state and return the person to it, i.e., they conduct the person in their life. In turn, the person also influences the evolution of the surrounding world. The thought processes happening in the brain of a person, the perturbations and accompanying actions and entropy changes, are reflected in the condition of the world around as triggers, initiating events in it, seeking to return the system of the person-world to a condition of greater equilibrium. Quantum mechanics gives an unambiguous example of the influence of the thought process of the experimenter on the result of his or her experiment. Here it is claimed that supervision over the quantum-mechanical object during the experiment changes the condition of the object. Therefore, a repeated experiment with the same object has to yield another result, different from the first. This is one of the reasons for the performance of ratios of uncertainty of Heisenberg and of the disappearance of corpuscular and wave dualism in experiments on an interference, when the experimenter knows (at the corresponding setting of the experiment) through which of the two slits (used in the experiment) the elementary particle passes.

1. Observation of the Influence of Collective Consciousness for the Operation of Generators of Casual Noise. (Global Consciousness Project, Princeton, NJ.)

The observation of entropic interaction within the international scientific Global Consciousness Project [58] is an impressive example of the impact of entropic perturbations happening in any part of a thermodynamic system on its other parts. This project started in 1999. In its framework, data from the generators of causal signals (random event generators) are collected. The role of such signals is played by electronic noise (quantum-indeterminate electronic noise) made by these generators. The generators are in 70 different places, scattered all over the globe.

Their data were continuously registered and compared for more than 12 years. The result was sensational: if somewhere there were any catastrophic or high-emotional events, all of the generators showed a deviation of electronic noise from their casual character. It happened during earthquakes, eruptions of volcanoes, military conflicts, and an act of terrorism in New York, when the World Trade Center was destroyed and 3000 people were suddenly killed. Deviations of the signals from their random character take place in such cases. The deviations were recorded by computers and explained as an impact of the collective thought process (consciousness) on the structure of the surrounding space. (Numerous examples of mental interaction between people are described by Dean Radin in his monographs [59, 60]. An influence of the consciousness of people on the events occurring in the world surrounding us is described there as well.)

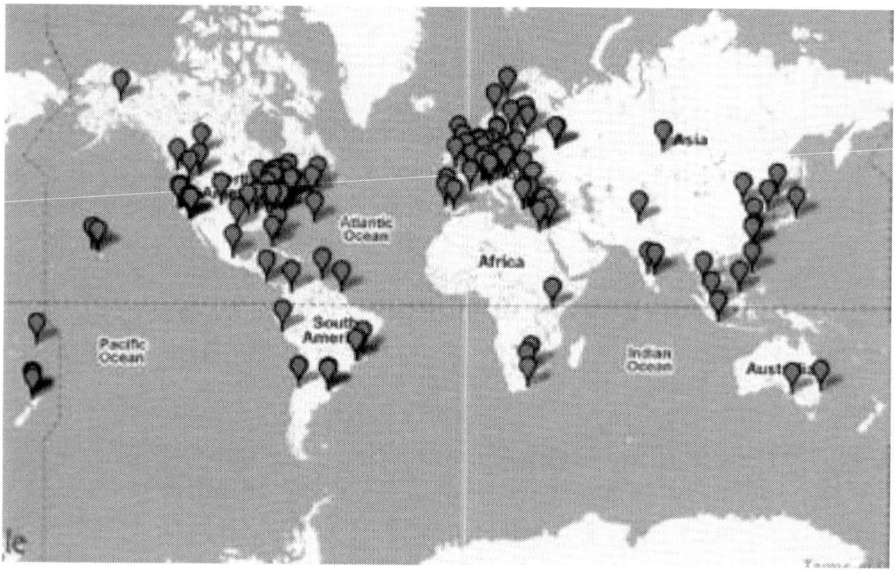

Figure 28. Distribution around the globe of the random event generators participating in the Global Consciousness Project, Princeton, N.J.

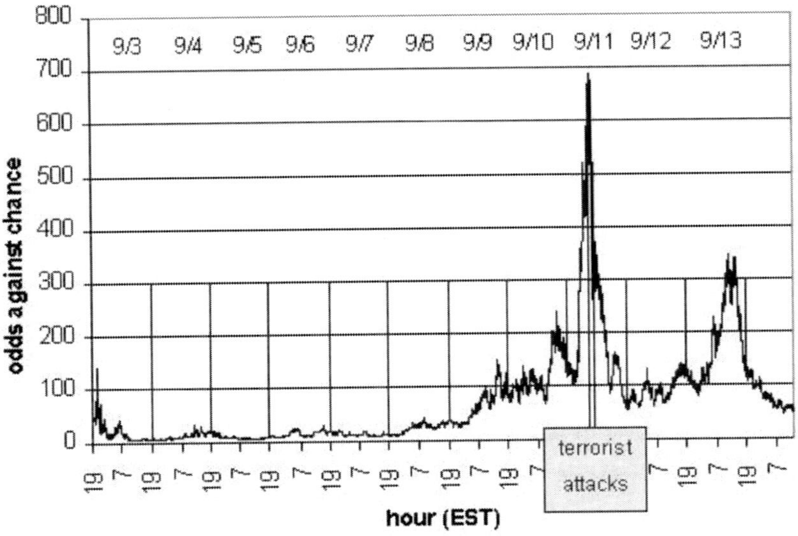

Figure 29. Splash fixing in the operation of generators of causal signals after an act of terrorism on September 11, 2001 in New York.

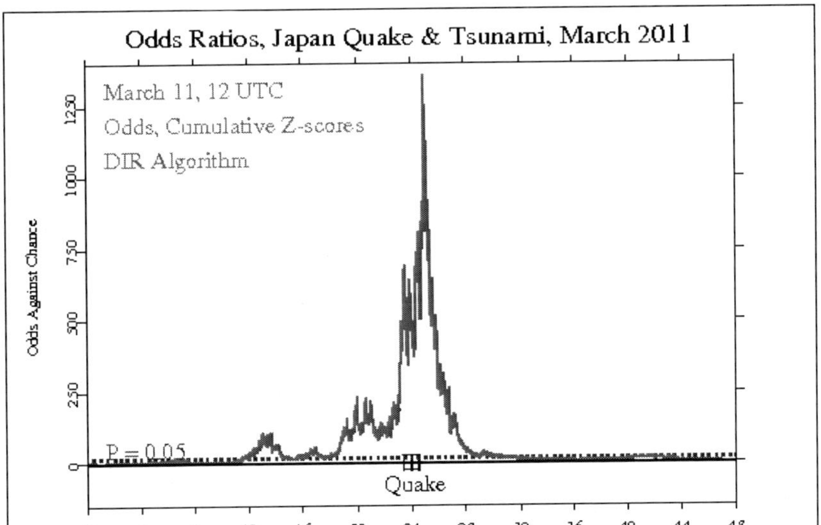

Figure 30. Splash fixing in the operation of generators of causal signals after an earthquake measuring 8.9 points and the subsequent tsunami on March 11, 2011 near the Japanese island of Honshu.

The figures show the sharp splash which occurred in the operation of the generators of random signals during an earthquake measuring 8.9 points and during the subsequent tsunami on March 11, 2011 at a distance of about 80 miles from the Japanese island of Honshu.

For more than 12 years during the observation within the project the similar behavior of the generators of causal signals was recorded during 429 events of catastrophic or high-emotional character (such as the opening of the Olympic Games, the finals of the World Cup in soccer, etc.)

2. Examples of the Influence of Thought Processes on Events

Before giving examples of the influence of thought processes on events, I want to draw attention to a remarkable custom which exists in the millennia, probably for all of the people on the Earth. It is a wish of health, good morning, good day or good evening at a meeting with friends or even with strangers. "Zdravstvuyte," Russian people say. (That means "be healthy".) "Sholom Aleichem," Jewish people say (which means "Peace for you"). People with any languages say to each other: "Good morning!" "Good afternoon!" and "Good evening!" The same wishes are heard in parting too: "Have a good one," "Good Luck!" "Take it easy," etc. And it is magnificent! Because, without thinking about it, we are increasing the probability of favorable events happening to those to whom we address these words (certainly, provided that they are sincere).Now we will consider a few purely speculative examples to show how entropic perturbations in our brain influence events.

1) For example, someone who is outside a train wants to stop the train at a particular spot. To make this happen, a chain of events must occur, resulting in, for example, someone on the train jerking a stop-cock, or (if it is an electric train) turning off the electricity supply, or the driver deciding to stop the train when he sees an obstacle on the tracks. That is, the chain of events may be different, but the result will be the stopping of the train. A distant analogy to this can be seen in quantum mechanics, where the system can move from State 1 to State 2 in different ways, but the probability summed over all possible paths will be appropriate to moving from State 1 to State 2.

2) An example from the field of sports: We know that sports teams show better results in their own stadia, but not during away games. This is true for all sports and is well confirmed statistically. Of course, for a home game, you

can prepare better; there is no discomfort associated with moving to another city, and each team wants to “avoid losing face” in front of their fans, etc. However, entropic interactions between fans and the team play an important role in the results. As already noted, the team has put themselves in the mind set to give a good performance in front of their fans. And the fans, in turn, are eager to see a good, winning game from their team. The entropy of a crowded stadium is drastically reduced. The team-viewers system is essentially taken out of a state of equilibrium, and to return to that state, the game, demonstrated by the team, should meet the expectations of the many spectators, and of the players. Only then will the tension at the stadium drop and return to its normal routine. In other words, for the players of this team, everything has to turn out as well as possible. The host team has to win. Only after that will the tension at the stadium fall and return to the usual track. And you must not forget about the probabilistic nature of the events. The probability of the victory is higher on the home field than on the opponents’ field, but it is not 100%.

3) A similar influence of thought processes on events happens when believers gather for a joint public prayer to ask God about something very important for them; for example, asking for rain in a drought. Sometimes it happens that their request is satisfied, and their belief is realized. In the system formed by this group of believers and the Nature surrounding them, the entropy is lowered, the balance is broken, and for its restoration, there has to be a chain of events in Nature, as a result of which clouds appear and it will rain.

In summer 2010, when the surrounding area of Houston was threatened by a rapidly looming hurricane from the Atlantic, one of the TV channels showed a pastor (a black woman) leading parishioners to the ocean shore, where they raised their hands toward heaven and prayed for the storm to pass beyond them – and it happened. The hurricane strayed a bit and hit the coast a couple of dozen miles away. In the system comprised of the congregation and the adjacent space, including the area covered by the hurricane, there was a certain chain of events that led to a change in atmospheric pressure and, consequently, to a change in the direction of the hurricane. This event can be considered similar to the previous examples, where the entropy of the interaction of humans with Nature, and their faith in their ability to influence events in such a way, acted as a trigger and was implemented through a chain of events in the natural environment.

It can be said that religion is of vital importance, because from generation to generation, people are convinced that it may help them in difficult situations.

4) The novel by the American writer Thornton Wilder called "The Bridge of San Luis Rey" gives us an example of an interesting analysis of the events which occur around us. The novel won the Pulitzer Prize in 1928. The events described there occurred in Peru in the eighteenth century. A suspension bridge collapsed between the cities of Lima and Cusco, leading to the death of five people who had been on it. The narration is led by a certain Franciscan monk who tried to understand whether the deaths of these people were casual or whether "God's idea was behind it." His investigation led to the conclusion that all of the victims had executed their vital mission and desired a resettlement to "The Other World." The "entropic approach" to the treatment of the events leads to a similar result, but replacing "God's idea" with an entropic perturbation in the consciousness of each of the five victims and, probably, in the consciousness of some people closely known to them. This perturbation imprinted a thought that they had reached everything in their life to which they had aspired, had done enough good deeds and could with a clear conscience "go to have a rest." From the moment of the emergence of this thought, this whole group of people was removed from a state of equilibrium. Their entropy went down, free energy increased and for the restoration of the lost balance there had to be a chain of events which came to an end with the bridge crashing down.

5) Now for some examples from the field of battle: Imagine two armies (approximately equal in strength) coming together at a decisive battle. Who will win? In accordance with the principle of least action and its interpretation used here, it would be the army whose will to win was greater, whose commander had a greater strength of will, and a greater belief in their invincibility. Remember the Battle of Poltava, described by the best Russian poet Alexander Pushkin:

"And suddenly, deafening the field,
Far burst: "Hurrah!" -
Regiments saw Peter.
And he flew before regiments
Powerful and joyful, as a fight.
He devoured the field with his eyes.
Behind him rushed a crowd
Of baby birds of Peter's nest
Within a terrestrial fortune,
In governmental and war deals
His companions, sons...

And before blue ranks
Of his aggressive teams,
Borne by faithful servants
In a rocking chair, being pale, immovable,
Having a wound, Karl appeared.
The confused look represented
Extraordinary worriment,
It seemed, the future battle
Has brought Karl into bewilderment.
Suddenly, by a slight wave of the hand
He sent regiments to attack the Russians."

The Swedish King Karl, on the eve of a decisive battle, was wounded. He was weak in spirit. He was not sure of the outcome of the battle. On the other hand, the Russian Tsar Peter was in full power and desired to win. He was confident of his victory and ignited his army. The Swedes, who had not known any defeats, were defeated by Peter's army, which was fighting for the first time!

6) Napoleon was an excellent military leader. He believed in his star, and won all of the battles in which he participated, but he lost the decisive battle at Waterloo. He caught a cold, felt bad and decided to get some sleep before the battle, asking his men to wake him up at the right time. However, his marshals did not wake him. They decided to excel and win the battle without him. It did not work. They did not have the leadership talent of Napoleon. Their spirit of victory was less than that of the enemy. When Napoleon awoke, it was too late to correct the situation.

In the conclusion of this section it should be underlined once again that changes in entropy influence the probability of the state of the thermodynamic system. This probability is never 100%. Therefore, the desirable change of a state, the desirable event, is not necessarily inevitable, but it will happen with a certain probability. It is possible to expect that in a drought the rain will fall without saying any prayers about it. But the probability of this event is small. A public prayer for rain increases this probability, let us say, from two percent to twenty. But this 20% still does not mean that the long-awaited rain will shortly spill forth. It is necessary to always remember this probabilistic realization of entropic changes.

3. Entity - John of God (Number 7 from "The Magnificent Seven")

(Entity - the original personality, the person being held in deep respect and possessing quite unusual qualities.)

In the examples described here, a lot of attention has been given to the influence of the collective thought process (consciousness) on occurring events. I personally faced this phenomenon unexpectedly, having visited Brazil in 2005. That year I had heard from a New York journalist, David Guy, about an outstanding Brazilian medium who, in the English-speaking world, is called John of God.

David and his colleague Igor Shikhman visited John and then wrote a book about him called "Working Wonders" [61], which intrigues us with surprising facts. The book (which I will not retell, although I strongly recommend reading it) describes in detail the biography of the medium, the methods used by him for the treatment of patients, and numerous examples of the healings he has performed.

Figure 31. "I am the happiest person in the world as God charged me with the mission." John of God - Joao Teyheyra de Faria.

Soon after the meeting with David Guy, one of my Boston friends, Tolya Zaitcev, told me that his close acquaintance (living in New York) had recovered (by means of John's treatment) from multiple sclerosis, against which modern medicine is powerless. Previously, he had been unable to leave his wheelchair, but after having visited John several times, he was easily able to reach his New York apartment on the 3rd floor, without using an elevator. All of this information sounded very convincing, and as I mentioned previously, for some time I had been trying not to miss any opportunity to become acquainted with people possessing unusual abilities. Anna and I decided to go to Brazil during our next vacation. I will admit in confidence that I was also attracted by an opportunity to visit the legendary Maracanã stadium to see a game with the Brazilian football virtuosos themselves and to communicate with their fans.

Rio de Janeiro surprised us by the harmony between the modern big city and the hilly ocean landscape, in which it is picturesquely built. The coastline of Rio with Copacabana beach, Sugarloaf Mountain and the Christ the Redeemer statue is so beautifully and architecturally formed that it is a World Heritage monument. The windows of our hotel room looked directly toward Copacabana. The beach was never empty. From morning to late evening it was filled, not so much by vacationers under the summer sun, as by numerous sports groups playing soccer, volleyball, practicing yoga, acrobatics, rhythmic gymnastics and other sports unfamiliar to us. And here, on Copacabana and in Maracanã and simply in the streets we were shocked by the genuine love of the locals for the city, for the country, for the football players, for the customs and for the numerous mediums. Statistically, in Brazil there are many more mediums than in any other country. People with religious fanaticism trust them and with a child's spontaneity enthusiastically they tell stories of wonderful recoveries due to their help.

From the viewpoint of the entropic approach to healing, the collective consciousness of the Brazilians, which has been adjusted this way, certainly helps healers and promotes their medical process. Moving forward, I will say that there is always a group of mediums of a lower rank near John of God during his healing sessions, helping him by efforts of their will.

One of the Entity's assistants translates the information which this patient shares into Portuguese. Behind the patient, the line of the next patients is visible.

Along the wall by the windows sit the mediums, who are helping the Entity through their strong-willed efforts.

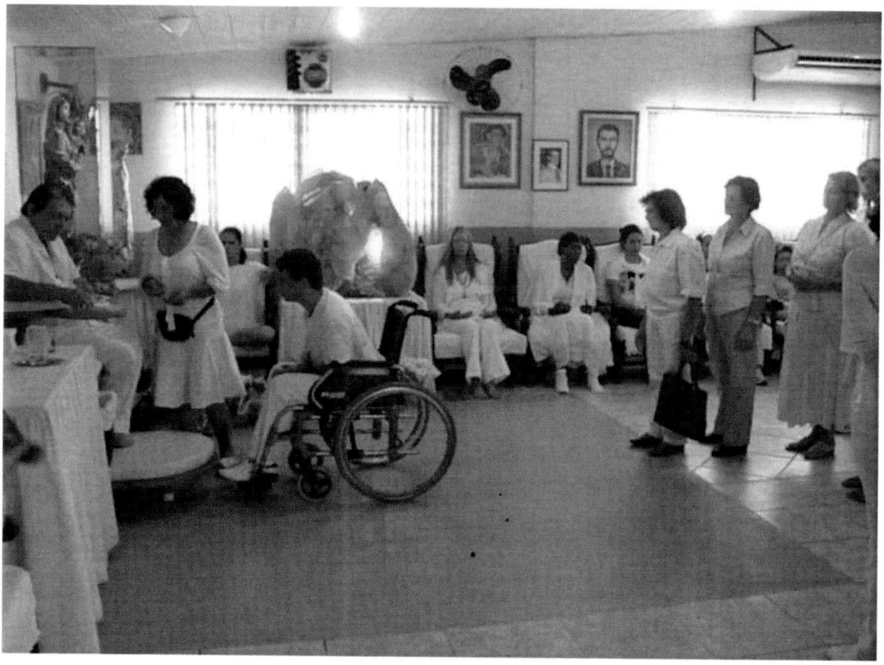

Figure 32. Entity accepts the patient who is in a wheelchair.

The original name of John of God is Joao Teyheyra de Faria. He is respectfully called Entity, meaning a personality, and given a capital letter to show his status, as having unusual gifts and abilities. Joao has accepted patients (or, rather, pilgrims) for about forty years in the small rural town of Abadiania, located approximately two hours' journey from the capital of Brazil. Visitors from all over the world come here to undergo treatment, or simply to communicate with Joao. Among other people, Joao has been visited, for example, by the spiritual leader of Tibet, the Dalai Lama, who afterwards left a letter of thanks, now hanging up for public inspection. And recently, the popular singer Lady Gaga visited Joao as well. He refuses nobody, he does not take money from anybody, and he personally pays for the daily meals (made to his recipes) for all the visitors. However, donations from the visitors are not forbidden. Joao considers that he is carrying out a mission assigned to him by the Lord and does not stint on either the money or the time required for its performance. So, on one of the days of our short stay in Abadiania, a bus carrying pilgrims from a remote region of Brazil was expected. The bus was expected to arrive in the afternoon, but for some reason was late and arrived long after midnight. Joao waited for the bus and accepted each of the pilgrims, and only at sunrise of the next day did he leave for home (he lives with his family on a farm, approximately 20 kilometers from Abadiania).

In the fine book "Working wonders" which I previously recommended for reading, there are many examples of the wonderful healings performed by Entity. We saw the museum room filled with wheelchairs and crutches which people, having been cured, had left there as superfluous. We became acquainted with certain visitors and whole families who were visiting not for the first time, but were regularly coming for communication with Entity. The stories they told struck our imagination. I will dare to give only one example characterizing Joao not only as a healer, but also as a person, who really devoted himself and his unusual abilities to the service of the people.

In Abadiania we stayed at a small hotel, whose manageress we had contacted before leaving Boston. The history of this attractive woman (about fifty years old) is very interesting. At the age of about 30 the doctors had found an inoperable cancerous tumor in her brain. Her days were numbered. Somehow she had learned about John of God. Being a person of action, she immediately went to Abadiania. The journey was not short because this woman had lived in Australia. Entity undertook her treatment, and in two weeks she felt an obvious improvement of her state. However, the tumor did not disappear fully. Entity said that for full healing she would have to stay nearby for a long time. And that is exactly what she did, although it required her to get divorced from her husband (who refused to move), and to leave her relatives and everything that she had known on the continent far from Brazil.

In Abadiania she continued to be treated by Entity and lived on the modest income she earned from cooking dinners for visitors. After a while (I do not remember exactly; a year or two) her persistence and diligence, fed by the attention of Entity, bore fruit. She had a small lodging built for herself, began to accept guests, and then opened a hotel for ten people. Now, she has a small farm near Abadiania, the products of which (vegetables, fruit, meat and dairy) she uses to feed her lodgers. She has recovered and is quite happy with her destiny. From time to time Entity gives her important tasks. He is often asked for help by the inhabitants of very poor and unsuccessful regions of Brazil, and he refuses nobody (and grateful people do not refuse him either). From time to time he has gone to his former patient, given her a plane ticket (bought by him) and asked her to go to a suffering area and to make something to raise their standard of living. And she has gone there, lived there for some months and organized, at her discretion, some business suitable for the situation. As a rule (do not forget about the ability of the Entity to influence events) these projects have been successful. The lives of the locals have been improved, and the envoy of the Entity has come back to her daily household chores.

The Entity has about a dozen volunteer assistants. They organize visitors, give them lectures, help on various councils, and serve as translators – all voluntarily, without any payment. Moreover, the Entity forbids them to accept any gifts, and violators are simply expelled from Abadiania. All of the assistants to the Entity are people whom he once helped, having once cured them or their relatives. Ingratitude, they have remained in Abadiania and devoted themselves to the Entity's service. You may ask: "What do they live on?" The answer is simple. They have started small businesses in Abadiania: coffee houses, cafeterias, fitness clubs, little souvenir shops and so forth. Some have businesses outside Abadiania and manage them from a distance. And all of them were successful, as was the owner of our hotel!

Like Tatiana Trofimova, Entity practices treatment using water he has structured. We brought a few bottles of such water back from Abadiania, froze it and photographed the water in one of them, using a microscope. In the photograph it is visible that this water is also polarized, like Tatiana's water (balls of stiffened air are extended and focused in some primary direction).

Around the Entity there is a lot of religious mysticism. In the estate yard where he accepts sufferers, there is a podium with a wooden "Triangle of desires." There is always a small line of people near the Triangle. One can leave a note on the base of the Triangle with a cherished desire. All of these people leave their notes, wait and hope for the execution of their wishes.

Now I think that my mission may be the understanding of the quintessence of a number of important laws of Nature, which came to me after long years of scientific activity. It also includes the writing of this book, which I had not thought about at all during my visit to Abadiania.

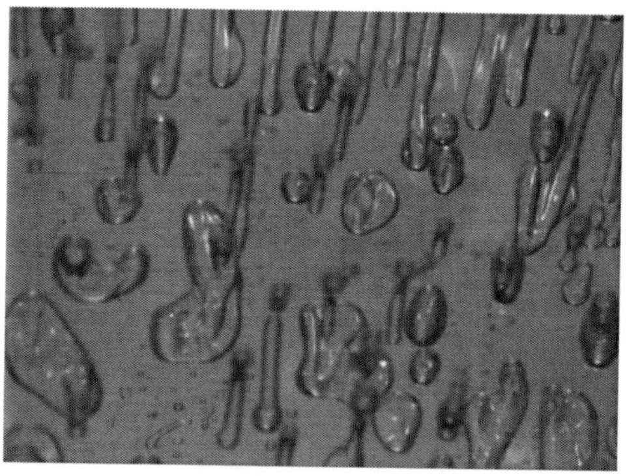

Figure 33. Water which was structured by the Entity.

Leaving a note in the "Triangle of desires," I was guided not at all by religious or mystical reasons; nor am I guided by them now, when writing these lines.- As I already tried to prove scientifically, I believe that the consciousness of certain people and the collective consciousness of some of them, can influence the surrounding Nature and events occurring in it. Thousands of visitors to this Triangle departed from it with a hope. The corresponding entropic perturbation in their consciousness smoldered like an unquenchable candle and also left a print not only on their course of life, but also on this prayerful Triangle. When you are approaching the Triangle, you feel some calm and a relaxation in your body, i.e., everyone becomes a part of the thermodynamic system uniting him or her with this Triangle, and with its past and future visitors. The entropy of this system is lowered, and your organism tries to lift it, increasing your entropy. It results in an observed effect of rest. We feel something similar when visiting churches, synagogues, mosques and other religious temples. Leaving our requests there in some or other form, we try to influence events in this way.

Figure 34. "Triangle of desires" with which people connect the execution of their desires, contained in their notes.
Anna and I also left notes. I wrote that I would like my vital mission to be realized. What that mission specifically consists of, I didn't know, of course. But I believed that it existed, just as a mission exists for each of us. Remember the reasonings on this subject by the scientific monk, Thornton Wilder, in the book "The Bridge of San Luis Rey", which I described in one of the previous sections?

Figure 35. One fan of the Entity entropically interacts with "A Triangle of desires," having previously put in it a note containing a cherished desire.

From this point of view, the Western Wall in Jerusalem is very indicative. Tens of thousands of people leave notes daily with their cherished desires between its stones.

Figure 36. Visitors entropically interacting with the Western Wall in Jerusalem.One gets a similar feeling when visiting the Temple of Athena on top of the Acropolis in Greece.

The atmosphere during the Entity's receptions is also saturated with religious mysticism. Appropriate lighting is chosen, and quiet, relaxing music is played. (Pay attention: during classes in the method of Reiki an identical situation is created.) Before the visitors appear before Joao, people collectively pray and meditate, directed by his assistants. Anna and I also underwent this procedure, and then I asked Joao to work on the cataracts in my eyes. I had decided in this way to conduct an experiment on myself to become acquainted with the feelings during such an unusual treatment and to check its efficacy. Joao wrote out for me a certain prescription, which asked me to return to him two days later for "a surgery," and before that to prepare myself by taking two quartz "baths" and some tablets from a local drugstore. During this meeting I did not feel anything special, but under the influence of the quartz "bath" my whole body noticeably pulsed.

Figure 37. Temple of Athena on top of the Acropolis in Greece.

On the appointed day I came to the "surgery," having previously undergone the meditative procedure described above. I was put in a chair which stood on a podium. The hall, in which there were about three hundred people, became silent, waiting for a miracle. The Entity opened a package with a flat sterile spoon and carefully scratched my right eye with it. All of this lasted no longer than a minute. I expected its repetition for the left eye, as my cataracts affected both eyes equally. However this did not happen, but the manipulation with the right eye had to act on the left eye too. During the "surgery" the Entity (as his assistants explained to me) was in a condition of a trance, and in his imagination his hand was being operated by one of the best-known Brazilian doctors from the past. I had a feeling of some weakness after the "surgery." I was recommended not to get up within 24 hours, and when in due time I got up and attempted to walk, I still felt uncomfortable. According to the received instruction, a week later I tried to read. It seemed to me that more light was getting to my eyes than had previously been the case, and the haze through which I saw a text before the "operation" had disappeared. However, some months later it appeared again, and my ophthalmologist continued to see the presence of cataracts in both of my eyes. It was difficult to say whether the cataract had decreased or not. After that, five years passed without any notable changes in the cataracts, and I want to believe that, by his touch, John of God, although not removing them, suspended their further development. When we were due to leave Abadiania, Anna and I again went to the reception to thank the Entity and to say goodbye to him. He took Anna's hand, traditionally wished her health and everything good, and told us that he will continue to always care for us. Further, the mysticism began. Within two hours we were in a taxi and heard radio information about the disappearance (connected with an accident) of a passenger liner flying to Europe (the remains of which were only found in the ocean a year later). Around us a thunderstorm roared. Grey clouds and pouring rain were seen at a distance of several hundred meters on both sides of the road, ahead and behind us, but over us there was a clear sky.

We were soon near the airport. Clouds hung over it, and we worried that our flight would be delayed. To our delight, when we arrived at the airport about 10 minutes later, the sky above it had cleared. We had bought tickets in Boston, but here at the check-in desk some hitch appeared, although the problem was not clear to us. Language difficulties prevented us and the airport employee who was checking us in from resolving the problem. But an elegantly dressed man standing behind us in the check-in queue spoke to us. In good English he offered his help and with our consent began to conduct negotiations in Portuguese with the employee serving us. In a minute or two the problem was settled. When our voluntary assistant was also checked in, he approached us, and we got talking. Having learned that we had been in Abadiania with the Entity, he told us that he knew him well and was even a distant relative. Moreover, last summer they had seen each other at an annual family gathering. He lived near Boston and came to Brazil on business. As there was some time before we needed to board the plane, he kindly recommended that we go up to the second floor of the building to admire the view of the city. We did this, remembering and discussing the last words the Entity had told us at parting.

Later, when we were sitting in our seats on the plane, we once again saw our well-wisher. He approached us to ask if everything was ok with us, treated us to ice cream and said that we could count on him if we needed anything.

VII. Correlation of the Entropy Changes and Rhythm of Time in the Processes Happening in the Organic World

We will consider what consequences occur in the organic world when the rhythm of the internal time in the isolated thermodynamic system is proportional to the speed of its increasing entropy. Let us start with an egg cell. The process of fertilization is a kind of perturbation of the state of an egg cell. After fertilization, its structure changes significantly, becoming more complex. The entropy decreases. The thermodynamic equilibrium is disrupted. To realize the recovery of the cell state, some biochemical processes start. They are seeking to increase the entropy of the cell and provide it with energy. During these processes a single-cell structure develops into a multicellular structure, increasing its entropy, although the state of each cell of this structure is disturbed, i.e., has a low entropy, compared to the original state of the unfertilized cells. As a result, the entropy of this formation is continuously increasing. As a countably additive function, the entropy of the multicellular structures is summarized from the entropy of each cell and increases, respectively, faster than at any subsequent time compared to the previous one, since the number of cells in the developing structure is continuously increasing, and the rate of growth initially also increases accordingly. In accordance with the proportionality of the rate of internal time and the rate of entropic growth, at this time interval the rhythm of the internal time also increases in this system. This will continue until such time when the number of newly emerging cells is equal to the number of dying cells, and then becomes smaller. Over time, fewer new cells will appear, and more and more of the old cells will die. This update is characteristic of the cellular composition of every developing living organism. Accordingly, the entropy of the body at first grows slowly relative to the standard entropy, and then more rapidly; at some point it slows down its growth, and then it completely stops. In correlation with the change in entropy, the rhythm of the internal system time after an initial period of rapid growth relative to the standard (external) time will slow down, and then will become equal to it; that is, its growth relative to the external time stops. At this moment, the life of the living body ceases. Graphically, the change in entropy and thus the internal time of a living system is an S-shaped function of the external time. Each of these functions has an inflection point corresponding to the maximum rate of entropy increase and the maximum internal rhythm of time. A straight line on the graph corresponds to the dependences for the inorganic system.

Figure 38 illustrates this dependence. The entropy changes along the ordinate axis. Along the abscissa axis external (standard) time changes. The speed of the change of entropy is set by a tangent tilt angle in each point of the plot. It is equal to the corresponding standard values in an initial and final time point and it has a maximum at an inflection point.

In Figure 39 the internal time changes along the ordinate axis. Along the abscissa axis external (standard) time changes. The rhythm of internal time is set by a tangent tilt angle in each point of the plot. It is equal to the corresponding standard values in an initial and final time point and is maximum in the inflection point.

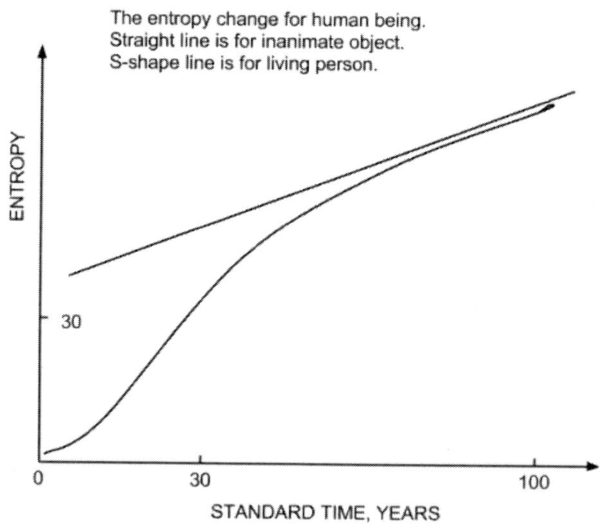

Figure 38. Change in entropy of the person with age.

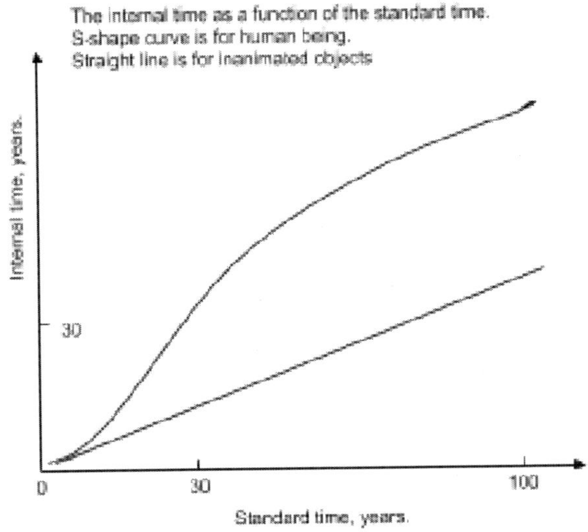

Figure 39. Change in the person's internal time with age as a function of the standard time.

For a multicellular system (e.g., the human body), it is possible to make the following conclusions from the plots of Figures 38 and 39:

1) The thermodynamic state of the embryo (fetus) is characterized by some value of the entropy associated with its life. In this case, the value of entropy can be smaller than the entropy of the corresponding non-living systems, i.e., a model of the embryo.
2) With the development of the embryo, its entropy increases, first slowly and then faster and faster. The internal rhythm of time behaves accordingly.
3) The rate of entropic increase and the internal rhythm of time reach a maximum value at the inflection point, and then begin to subside. By this time, the number of emerging cells becomes equal to the number of dying ones. The embryo has passed through the stages of a child, teenager, and young person. It has reached adulthood; it has become an adult.
4) After reaching adulthood, the growth rate of entropic increase is slowing, and the internal rhythm of time reaches a maximum value, and begins to slow. At the inflection point, the person gets the fastest reaction within his or her genetic (physical and mental) capabilities.

During this time, he or she sets world records in sports, and reaches the peak of his or her mental abilities and activities.

5) On the other side of the inflection point, the entropy increase gradually slows down, and with it, the rhythm of the internal time slows down as well. High-speed physical and mental abilities are declining. The body is aging. When the graphs of Figures 38 and 39 become parallel to the graph for the inanimate systems, the entropic increase associated with the life stops. The internal time ceases to differ from the external time, and the process of activity stops. The person dies and his or her organism (the body) transfers from the world of living Nature to the world of inanimate Nature.

The considered age dependences of the change of entropy and the corresponding rhythm of time have to be manifested at long relationships between people of different ages. This relates to married couples, children and their parents, and school and university students and their teachers. Each of these groups forms a thermodynamic system. Its separate parts interact entropically, seeking to make the distribution of entropy in the system more uniform and thus to increase the entropy of the system, having at the same time accelerated all biochemical processes proceeding in it. As a result, there is a rejuvenation of organisms in the participants of each group (this does not relate to little children who are in an initial stage of slow growth of their entropy). At the same time, all younger people experience faster development and mature. As for unevenly-aged spouses, the effect of the rejuvenation of seniors was confirmed with the researches conducted and published in 2013 at Stanford University under the leadership of Professor Brook. So it is not necessary to be wary of marriages unequal in age; one should pay more attention to children and remember that they, as well as the noble work of school teachers and university professors, are generously paid by Nature.

VIII. Using Changes in the Internal Time of a Living System to Slow Down the Aging Process

Understanding the correlation between the increase in the entropy of a thermodynamic system and the rhythm of its time can positively intervene in the aging process, slowing it down. It suffices to find appropriate mechanisms for the impact on the entropic state of the system. When applied to a person, some of these mechanisms are already in use, but without the involvement of entropic interactions for the explanation of the results. One of these mechanisms is moderate physical exercises. Each muscle is structured and loses entropy at any physical tension, and the whole body loses entropy together with its muscles. This decrease in entropy, algebraically adding up to the common standard entropic increase over time, leads to a decrease in the rate of growth of the entropy of the whole organism, i.e., to slowing down the rate of its internal time. The more muscles are involved in this process, the greater the overall effect. The muscles get tired and contract more slowly after a certain activity. All body movements are slow as well, and the slow rate of the internal time is the explanation for this. After a rest period, the muscles return to a state close to equilibrium. However, with regular physical activity, the muscles remain in tonus even between periods of exercise. That is, they do not return to the initial relaxed state. Their entropy is not fully restored, and the rhythm of the internal time lags behind the rhythm of time at the complete absence of exercise. If exercise is done regularly, not excessively, and if it does not injure the muscles, the total delay of the internal time becomes significant and slows the aging process.

Since the decrease in entropy in any part of the body leads to its reduction for the entire organism and, therefore, to the slowing of its internal time, regular non-excessive tensions of the brain and of the central nervous system also lead to the slowing down of the aging process. A correct improving massage and appropriate methods of manual therapy give similar results.

The entropy of the human body lowers and the rhythm of the internal time slows down in the case of sickness. Simultaneously, the patient loses energy. His entire muscular system recreates (relaxes). Due to this, the body increases the entropy, and the total entropy effect may be different from the entropic reduction prior to its enlargement. During recovery, the entropy of the diseased organs increases. This leads to the more rapid growth of the entropy of the whole organism and, consequently, to acceleration of the rate of internal time. This acceleration not only compensates for the slowing rate of the internal time during the illness, but it also covers the cost of losing muscle tonus. Thus, the disease usually accelerates the aging process.

On the other hand, living “away from the bustle of the city” somewhere in the open air or in a resort area allows you to relax, rest, avoid the “entropic shocks” of everyday urban life, and thereby reduce the rate of internal time, compared with that of the city. How does it work? The person and the surrounding physical space constitute a kind of thermodynamic system. The lack of significant perturbations of entropy leads to the coincidence of the internal time of the individual with an external (standard) rhythm, which corresponds to the life “away from the bustle of the city”. In the tense atmosphere of the city, full of different events, conflicts and stress management, the entropy of the surrounding space, as well as the entropy of the citizens is reduced. The thermodynamic equilibrium is disturbed. According to Le Chatelier’s principle, Nature opposes these changes. It seeks to return citizens and the whole city to a condition of greater equilibrium. Biochemical processes are initiated in their organisms, increasing entropy. This increase in entropy, adding to its standard value, increases the rate of growth of the entropy of the body, and hence the corresponding rate of its internal time. As a result, urban people age faster than those living in a more relaxed atmosphere. Statistics confirm that a large proportion of the oldest people live in rural areas. The existence of entropic perturbations in space and their influence on the human body is becoming more acceptable to the understanding, if we recall that, according to modern concepts, the vacuum is not just “empty”. It is a material one, has an energy in the absence of any fields in it, and changes its geometry in the presence of a physical field. It is useful to recall that, according to Noether’s theorem [62], the homogeneity of the physical space with respect to spatial coordinates, its relative isotropy and homogeneity with respect to their time in the four-dimensional space of the Special Theory of Relativity lead to the laws of conservation of momentum, angular momentum and energy, respectively. That is, the laws of Nature, our health and livelihoods depend significantly on the structure of physical space around us, from the current “order” that has a place in it.

Here it is pertinent to remind ourselves of Schrödinger’s opinion [50], which was already quoted earlier. According to Schrödinger, the acceleration of the rate of increase of entropy leads to a more intensive vital process. It is natural to expect also a boomerang effect at a delay in entropic growth, i.e., staying in one thermodynamic system, with the environment and interaction with it, is essential for the processes proceeding in living organisms. In particular, the accommodation in climatic conditions with a different temperature is also reflected in them. In the equatorial and tropical areas the thermal exchange of living organisms with the environment is less intensive than, say, in a temperate climate. Heat developed in living organisms is radiated into the surrounding space at a slow rate. The reduction of entropy connected with it is lowered. For its restoration the organism needs to show less activity. Therefore, the biological rhythm of time is decreased there and all vital processes are slowed down. With other things being equal, biological time passes more slowly here and people live longer here than in a temperate climate.

The report of the international group of researchers published in 2015 in the USA in the Proceedings of the National Academy of Sciences was devoted to the distinction between biological (individual) and calendar (standard for our galaxy) time (Published online before print, July 6, 2015, doi:10.1073/pnas.1506264112, PNAS July 6, 2015). The results of the research show that some of the people practically did not grow older in the 12 years during which the research was conducted; other participants grew three years older biologically for each year they actually lived. Those whose biological ages considerably exceeded their calendar ones performed badly in tests which are usually given to elderly people, regarding their coordination of movements and logical thinking. Physical efforts were also more difficult for them.

It appeared that the biological ages of some people who were born 38 years ago come nearer to the age of retirement.954 people from a city in New Zealand were observed. All of them were born in 1972-73. The indications of 18 biological markers, when the participants in the experiment were 26, 32 and 38 years old, were fixed. Scientists were interested in the condition of the kidneys and liver, the level of cholesterol and other similar indicators.

The analysis of the received results shows that for 38 calendar years the biological ages of some participants in the research corresponded to only 28 years, while for others it came nearer to 60 years.

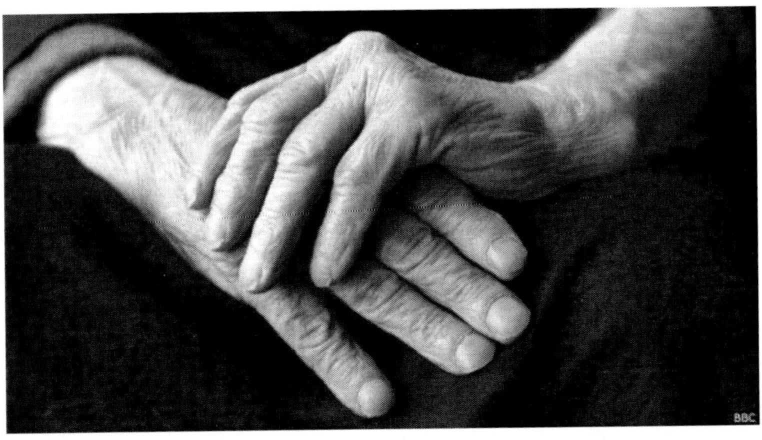

Figure 40. The hands of a 38-year-old participant in the experiment are shown in the photograph. Undoubtedly, they look like the hands of an old person (as with other characteristics of his health).

Biologically old people "look bad, in them there is no energy," states Professor Terrie Moffitt, who directs research from Duke University in the USA.

From the perspective of the thermodynamic approach, the different speeds of aging of the participants in the research are explained by their genetic features, their state of health, way of life, and the degrees of intensity of the situations surrounding them in terms of family, workplace, and circle of contacts. One's individual biological time significantly depends on the above-listed factors.

IX. Dependence of the Energy Characteristics of a Living Body on Temperature

Of particular interest is the comparison conducted by Dr. Vladimir Shpunt on the dependence of energy characteristics on the temperature for living (organic, i.e., biological) and inanimate (inorganic) systems. The human body operates in a temperature range approximately between 35 and 42 degrees Celsius. Its energizing occurs as a result of the assimilation of food intake. The biochemical reactions involved in this process lead to an increase in both the entropy and enthalpy components of the Gibbs free energy, which, in turn, tends to remain unchanged. The optimal temperature for a human body is 36.6°C. Outside the range of 35-42°C the activity of the human body actually stops and the temperature dependence of the enthalpy H and entropy s becomes close to linear and virtually identical to the corresponding dependence for the inorganic Nature.

After the digestion of food, the person spends the received energy on physical and mental activity, as well as on heat exchange with the environment. In accordance with the principle of least action, the entropy of the body decreases at the same time due to the structuring of what is happening in his or her muscle and cerebral cortex. The maximum of these changes, again, falls at the temperature 36.6°C. However, not all of the energy acquired in the metabolism (and therefore not all of the entropy) may increase the demand. Part of it remains in the body, accumulating in the emerging of new cells.

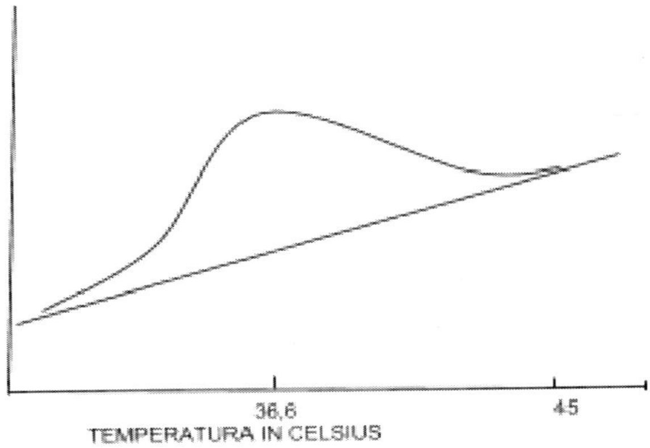

Figure 41. Behavior of internal energy of the person with a change of temperature.

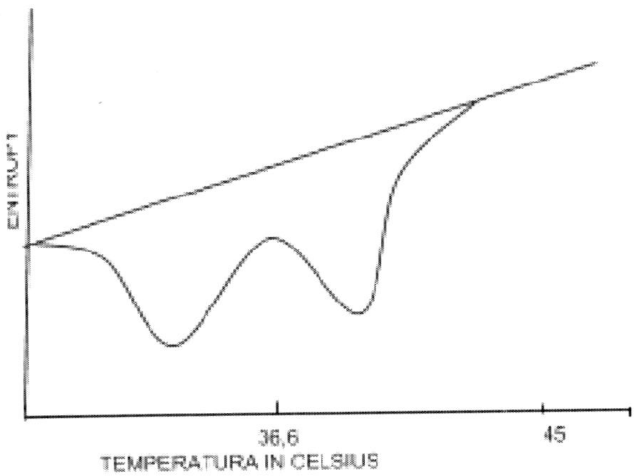

Figure 42. Behavior of entropy of the person with a change of temperature.

Outside of the range of 35-42°C values of enthalpy and entropy which are spent during the course of a life, become be a linear function of temperature as is in inanimate Nature.

Figures 41, 42 and 43 illustrates the behavior of internal energy of H, entropy of s and free energy of G at temperature change.

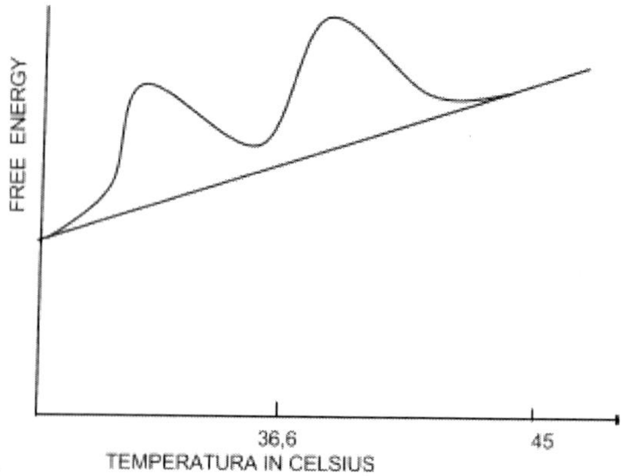

Figure 43. Behavior of free energy of the person with a change of temperature. ("Camel shape" in Shpunt's definition.)

X. Entropic Order Rules the World

"I trust neither camomiles, nor cuckoos
and to fortune tellers, you understand, I don't go"
Bulat Okudzhava

"Two things fill my mind with eternally new and increasing awe:
the starry sky over me, the moral law inside me."
Immanuel Kant

Really, the entropic order rules the World. It is appearing everywhere. We saw how this order works in thermodynamics, how the processes going on in the world of both living and inanimate Nature depend on it. In theories of elementary particles, such as "Standard Model" [63], "The Theory of Strings" [64], "The theory of electroweak interaction" [65] and others, all characteristics of elementary particles are defined by the relevant groups of symmetry, i.e. a peculiar mathematical description of a certain entropic order. The origin and further development of the Universe, the emergence and evolution of organic life on Earth, and possibly many other parts of the Universe as well are obliged to the change of entropy. It is possible to connect development of human society with it. Entropy plays an essential role in all aspects of life; it is closely connected with a rhythm of time destined for us.

The lines taken out in the epigraph fully reflect my relationship to superstitions and related fortune-telling according to cards, camomiles, coffee sediment and so forth. However, through astrological prediction an individual's destiny may be defined, which exists in the millennia and can be statistically justified with a probability considerably exceeding random guessing. According to astrology the character of the person, the opportunities that come to the person, his (her) mission and destiny in general depend on the arrangement of constellations - zodiac signs under which he (she) was born.

Certainly, a lot of things depend on casual events, on the circumstances meeting during the course of life, but finally these ways come on "a path" predicted by a horoscope, with surprisingly high probability, i.e., as a philosopher once wrote: "Necessity paves itself a way through a crowd of occasionalities". Such a success rate of horoscope predictions is scientifically inexplicable today, but it exists and makes us think, as well as a number of other phenomena lying outside today's scientific knowledge.

According to the scientific ideology stated in this book, our health, physical activity, thinking and a lot of other things are defined and directed by "entropic order". It is everywhere: in us, on Earth and in the sky. Look once again at the stars. They are myriad. They are scattered in the night sky, apparently chaotically. But close examination finds in their arrangement a certain order (The Milky Way, galaxies, constellations, etc.) which accompanies us all our life, from our birth till our last days. And as we already discussed earlier, this order isn't something abstract for us. In actual fact, it is generally defining our specific features. The structure of the World space, the global processes proceeding in it, and the fundamental laws of physics and of other sciences depend on the order. So, to each continuous symmetry of a physical system there corresponds some conservation law. Allow me to repeat once again that, according to the theorem of Noether [62], the uniformity of space leads to the law of the conservation of momentum, isotropy leads to the conservation law of the angular momentum, and uniformity in time, means that experiments performed at different times will yield repeatable results, leads to the law of energy conservation and so on.

Astrologers try to bind the fates of individuals and the events occurring in the world with an arrangement of stars and planets. The Ancient Chinese even used a board game called "GO" for this purpose. The Master of this game placed stones (as a bisque) on a playing board (these were figures used for the game of "GO", having the shape of lentils) in the form of a constellation under which the person expecting predictions of his destiny had been born. After that, the game between the Master and the client began. The order of an arrangement of stones, formed this way, allowed the Master to predict the future of the client. Over time the game of "GO" became the most popular in Japan. Today, no politician can hold a high public post in that country (for example, to become the prime minister) without being a Master of "GO".

When Immanuel Kant wrote the lines quoted in the epigraph: "Two things fill my mind with eternally new and increasing awe: the starry sky over me, the moral law inside me," he intended the term 'starry sky' to mean not only the beauty of the Universe, striking our imagination, but also the laws of natural science inherent in it, regulating our lives.

As for the moral law mentioned by Kant, this means the moral code that distinguishes human society from a primitive one. Everything begins with the birth of a child. It is possible to say that a child "absorbs this code together with the milk of its mother". No mother teaches her child bad deeds. The first books, animated cartoons, school education, and then serious literature, music, and art "sow in him (her) something reasonable, kind, eternal," i.e., things that will not "entropically disturb" society, by crushing the ethical standards accepted in it. These norms fix a specified "entropic order" which manages society and is written down in the form of the relevant legal laws.

It is also necessary to mention the undoubtedly positive role of religion in questions of moral education. Religion gave people a set of moral laws at the beginning of the formation of human society and supports it to this day, being its peculiar moral support. Creators of religious dogmas were able to bring to us through millennia the entropic perturbation which arose in their consciousness and was then reproduced by believers of all denominations on visiting religious temples, holy sites, at prayers and when reading sacred books.

In painting, music, literature, and the performing arts we easily find other ways of preserving and transferring entropic perturbations. The brush of the artist, the notes written down by the composer, and the lines that are left by a writer's or poet's pen transfer their emotional spirit at the time of creating the corresponding work. Centuries may pass, and many generations of people may be replaced, but the entropic perturbation recorded in these works of art, and the related entropic order in the consciousness of authors will be immortal, as well as the genetic code imprinted in molecules of DNA, or the structure of the starry sky over our heads!

XI. Application of the Thermodynamic Approach to Natural Phenomena

1. Scientific Search from the Perspective of the Thermodynamic Approach

"For the new theory to be correct, it has to be crazy enough."
Niels Bohr

The process of a scientific search is somewhat similar to the process of predicting the future. What do you do while trying to solve a scientific problem? First of all, you search through your memory trying to match the problem with your prior experiences. The next step is to search the available literature on the subject. If that doesn't help, you can discuss the problem with colleagues and experts in the field, set up appropriate experiments in order to resolve the situation, and so on. Meanwhile, your consciousness all the time is aimed at solving the specific task. Your entropy has decreased, and only after your mind finds the solution to that problem will your personal entropy increase, and you will return to your personal stable state. Many cases are recorded in the history of science, describing how a specific discovery was made by different people at almost the same time. Some examples are the law of Boyle-Mariotte [65], the law of Mendeleev-Clapeyron [66], and the law of Charles and Gay-Lussac [67]. According to the thermodynamic approach described here, I suggest that the entropic perturbation which occurred in Boyle's mind led to a similar "structural perturbation" in Mariotte's consciousness a few years later (when he started to solve the same scientific problem). This resulted in the emergence of a law of physics, named after both scientists. The same statement would be true about the other two above-mentioned examples.

Bearing in mind that the entropic information is independent on the size of the thermodynamic system – that is, that the information becomes instantly known to all of the parts of the system, irrespective of the distance between the parts – it is possible to propose the following: Somewhere, far away from our planet, possibly outside our galaxy, there are living organisms, inhabitants or aliens that have evolutionarily surpassed us, humans, in later stages of their development. Consequently, the laws of Nature that we are discovering nowadays were known to them much earlier. Some of us, thanks to our persistent and ever-searching minds, are able to successfully "connect," "login," or "hack" into the thermodynamic system, in common with these thinking beings. And when the mind is looking for answers to various scientific problems, it is not in a state of equilibrium. The mind has to obtain the information on a specific problem, and once it succeeds, our internal system, the consciousness, returns to an equilibrium state. "Eureka!" we exclaim, as the legendary Archimedes did, "I've been inspired – I discovered a new law of Nature!"

2. Faith of Mayan Civilization

A similar approach and theory would apply to the ancient Mayan civilization. The Mayan civilization was established in Central America about 4000 years ago and progressed to the highest level of their development during the first millennium, about 700 years before the discovery of America by Christopher Columbus. This civilization used hieroglyphic writing, discovered and established mathematical rules at a much higher level than their contemporaries in Europe, and possessed a deep knowledge of astronomy. It would be worth mentioning that the Mayan calendar was more correct and descriptive than the one we use today. But then, unexpectedly, for an unknown reason, this advanced civilization vanished. The cities the Mayans had built fell into decay. The rural economy, which had reached a high level of development, collapsed. The architecture of new constructions that had been built was later deemed primitive from an engineering perspective. Everything suggests that the most intellectual part of Mayan society had ceased to exist.

Today there are two accepted explanations for the disappearance of the Mayan civilization. One is related to a short-term drought that allegedly took place; the other blames a possible epidemic that wiped out the entire population. It seems to me that neither version stands up to criticism. Really, it seems very strange that the most educated and advanced society of people happened to fall victim to a natural disaster. Usually, things are reversed. Natural cataclysms, as a rule, lead to the collapse of the lower level of society, which lives in the worst economical and hygienic conditions and, therefore, would be less protected from natural disasters. Whatever happened to the Mayan civilization is mysterious and very intriguing.

Once, I travelled with my wife Anna and our older daughter, Masha, to Mexico on the Yucatan peninsula. We had read about the pyramids constructed by the Mayans, and wanted to see the remains of those personally and to compare them to the Egyptian pyramids. We rented a car and went to visit Chichen Itza. The poverty and wretchedness of the local settlements through which we passed on the way left a sad impression on our consciousness. Each time we stopped and got out of the car, we were surrounded by a flock of barefoot children begging for small change. Their parents were nearby, trying to sell some primitive artwork. Although we bought a few pieces of art and donated some change to the kids, it was very sad to realize that those poor people, living in poverty, were probably the closest descendants of what was once a great civilization.

In a few hours we arrived in Chichen Itza. It was a nice, sunny day. The town of the ancient Mayans was filled with tourists from all over the world. The biggest attraction was three remaining pyramids and ruins, forming a huge rectangle with four pillars. An essential difference between the pyramids of Chichen Itza and the Egyptian pyramids is that the Mayan pyramids are truncated at the top and are smaller in size. In addition, the Egyptian pyramids were designed as Pharaohs' tombs. The purpose of the Mayan pyramids and of their truncated shape remains mysterious. It is important to note that the design, structure and orientation of the Mayan pyramids is not random, but reflects astronomical events. The total number of steps on the sides of the biggest pyramid, the temple of Kukulkan, which is 30 meters high, is equal to the number of days in our calendar year, 365.

The balustrade aligns the sides of the pyramid. The exact mathematical calculation of the balustrade strikes the imagination. On the days of the spring and autumn equinoxes (on March 20 and September 22) at approximately three o'clock in the afternoon, beams of sunlight shine at its western part in such a way that the light and shade form an image of seven isosceles triangles, making the body of a thirty-seven-meter stone snake as though creeping (in the process of the sun's movement) to its own head, which has been cut out at the base of the pyramid.

We were fascinated with what we saw, and then finally smitten by the view of the "Holy Well" which we came across at the end of the day. It is located 300 meters to the north of the main buildings of the city and represents a giant truncated cylinder with a diameter of over 60 meters. Moreover, about a quarter of the well's cylinder had been chiseled off in a vertical manner, all the way down to the bottom of the camas, leaving it in a shape of an elongated letter D. The walls of the cylinder are steep and one gets the impression that they have been carved from a very hard rocky material. The distance to the water is about 20 meters, and the distance to the bottom of the well is probably the same again. In our opinion, the correct geometric shape of the well (truncated cylinder) speaks about its artificial origin (and not a natural one, as we are assured by a variety of sources).

Figure 44. Kukulkan's pyramid.

We went back to the hotel in Cancun late at night and were very excited by what we had seen. There were lots of unanswered questions. Why, in general, did the Mayans decide to build the pyramids and make them in such a truncated shape? What was the function originally intended for the well, assuming that it was built like a pyramid? What really happened to this civilization? How could it vanish in the absence of any obvious war and long before the arrival of Spanish conquistadors?

And could it be that the pyramids had served as watchtowers for observing the potential approach of an enemy? While some defensive armory was placed on top of the pyramids, there was also some equipment for astronomical observations. Reasonable, isn't it? Incidentally, one of the pyramids of Chichen Itza looks like a prototype of a modern observatory. As for the well, it could have been built as a source of fresh water, which was always lacking at the peninsula in the absence of rivers and lakes.

And what about the appearance, and then the mysterious disappearance, of the Mayan civilization?

Figure 45. Observatory ruins in Chichen Itza.

Figure 46. Holy Well.

The Mayans used twenty signs in mathematics with zero at the start, which looks more advanced than the one used at the time of the Old World.

Newton, Galileo and Copernicus were not yet born, but Mayan astronomers already knew all the necessary information that would allow them to calculate the solar calendar with a resolution of seconds.

The thermodynamic approach proposed in this book suggests the following. Among the Mayan people there were those who could be considered as "thoughtful clairvoyants". There may have been just a few of them or even only one. Like the other Mayan people, they believed in their gods, living somewhere in the sky. The Mayans would stare at them, begging for help in their daily lives. In the minds of those seers, entropic perturbations arose which forced them out of balance and reflected problems requiring a solution. The chances are slim, but they still exist, that those people could potentially "connect" to some reasonable (and far advanced in the knowledge of the laws of Nature) dwellers of a certain planetary system, different from the Earth, thus creating a mutual thermodynamic system. This system would be instantly taken out of the equilibrium state. In order to return it to the equilibrium state, tasks had to be resolved. The consciousness of the clairvoyant Mayans, who were trying to figure out a solution to a problem, came to know the information thus revealed to them, discovered the unknown and felt the decline of the entropic perturbations.

Earlier, we mentioned that such a hypothetical creation of a mutual thermodynamic system with the inhabitants of advanced extraterrestrial civilizations (far from our planet at a distance of many light years) would be the only real opportunity to establish contact and communication with them. Within one thermodynamic system, the entropic interaction and exchange of entropic information occur instantly, just as occurs in a system of entangled quantum objects. Therefore, the previously described telepathic contact would allow this communication to occur in real time without any delay. Clairvoyance will show us each other, including the world around us, just as it is accomplished in our own time using video over the internet. It sounds amazing, but it may be real!

In September 2014 the Journal "PLOS One" published a paper by scientists from Harvard University in collaboration with scientists from Spain and France. The paper states that human brains are capable of transmitting messages to one another without physical connectivity. They described an experiment in which a person located in India with a headset connected to the internet was able to transmit thought to the other three persons located in France, also with headsets connected to the internet. As reported in the paper, those four people did not use any of the standard communication methods such as texting, emailing, or talking. The authors said that when one of the people in the experiment thought of the word 'ciao,' meaning 'hello' in Italian, the other people had flashes of activity in their brains, as recorded by electroencephalograms. The four, as stated in the paper, were 4,000 miles apart during the experiment. Reportedly, this is the first successful experiment on telepathy.

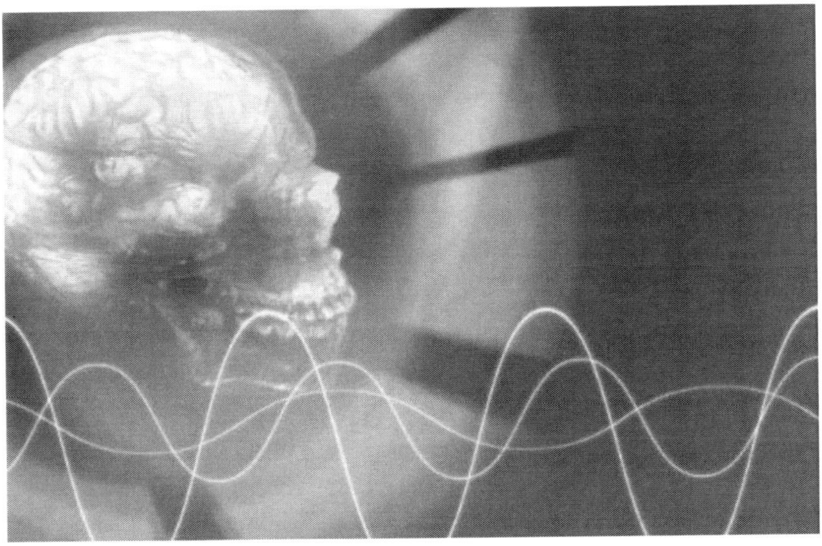

Figure 47. Harvard University scientists conduct successful telepathy experiment; 4 people communicate 4,000 miles apart.

Giulio Ruffini, a theoretical physicist and one of the co-authors, said: "It is kind of technological realization of the dream about telepathy, but it is definitely not magical. We are using technology to interact electromagnetically with the brain."

Researchers used internet-linked electroencephalogram (EEG), and robot-assisted and image-guided transcranial magnetic stimulation (TMS) technologies. Four healthy participants, aged 28 to 50, participated in the study. One of the four subjects was assigned to the brain-computer interface (BCI) branch and was the sender of the words; the other three were assigned to the computer-brain interface (CBI) branch of the experiments, where they received the messages and had to understand them.

In this first-of-its-kind study, an international team of neuroscientists and robotics engineers demonstrated the viability of direct brain-to-brain communication in humans. "We wanted to find out if one could communicate directly by reading out the brain activity from one person and injecting brain activity into the second person, and to do so across great physical distances by leveraging existing communication pathways," said Alvaro Pascual-Leone, of the Beth Israel Deaconess Medical Center (BIDMC) and Professor of Neurology at the Harvard Medical School.

Using EEG, the research team first translated the greetings 'hola' and 'ciao' into a binary code and then emailed the results from India to France. The computer-brain interface transmitted the message to the receiver's brain through non-invasive brain stimulation. The subjects experienced this as phosphenes, flashes of light in their peripheral vision. The light appeared in numerical sequences that enabled the receiver to decode the information in the message, and while the subjects did not report feeling anything, they correctly received the greetings.

The planetary systems closest to us are at a distance of 50,000 light years. Only there can one expect the existence of organic life similar to ours. Only there can one expect to discover reasonable inhabitants, for whose electromagnetic signals we are currently looking in outer space.

Did we do the right thing in spending billions of dollars in the quest for aliens? After all, can we expect that the evolution of the whole Universe initiated and continues to act in the same way as our Galaxy? It is conventionally accepted that the whole Universe appeared at the Big Bang. We have no specific reason to suspect that galaxies other than ours evolve at a higher rate than our galaxy. Yes, one can suggest that there are planetary systems whose development initiated 1,000 years prior to our Solar system, but not 50,000 years earlier! However, the outer space signals that we are receiving today had to be issued 50,000 years ago in order to reach Earth today. Yet, did humans have the capability of generating and sending such signals into outer space 50,000 years ago? Of course not! Therefore, the living organisms inhabiting other galaxies would not have been able to generate such signals 50,000 years ago either. Therefore, our expectations of receiving and recording signals sent from other systems 50,000 years ago are, to put it mildly, not justified.

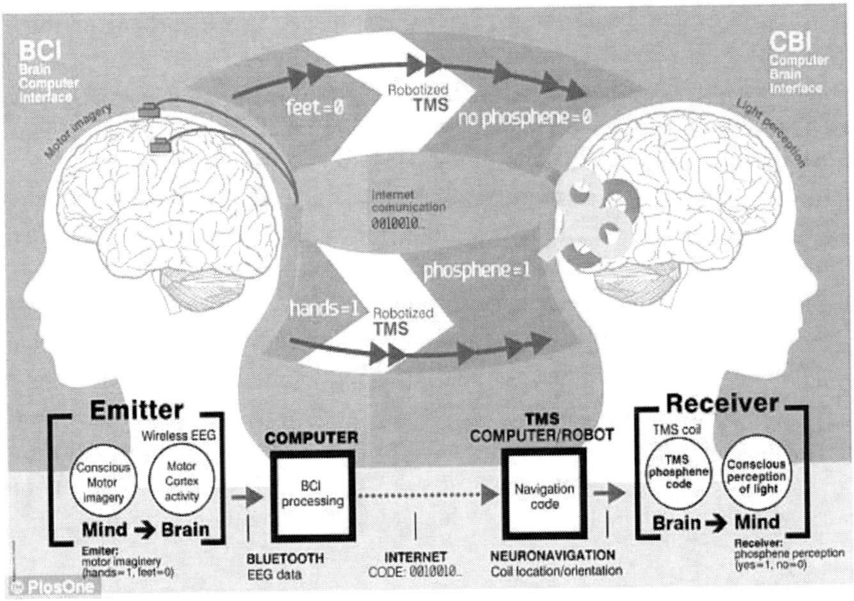

Figure 48. The computer-brain interface transmitted the message to the receiver's brain through non-invasive brain stimulation.

Equally unjustified are beliefs about aliens who can visit our planet, or that Unidentified Flying Objects (UFOs) are conducting surveillance of our lives. Even if there are machines that have capabilities of moving through space at a speed close to the speed of light, i.e., 300,000 km/sec, it would mean that by the time these aliens returned home, assuming that they wanted to, they would find themselves surrounded by their distant descendants. It is difficult to imagine who would want to experience that, and what meaning would be enclosed in such a journey. (Let us remind ourselves that with growth of the speed of movement, the rhythm of time is reduced, and at a speed equal to the velocity of light, time stops.)

However, from another perspective, it could be viewed as a time journey. While visiting Earth, these visitors would find themselves in the distant past of the Universe's intelligent life. And upon their return home, they would find themselves in the future. It is possible that someone could be interested in such a journey in time and would pursue the idea and dedicate their life to this. For me personally, it does not appeal! Mayan society highly respected clairvoyants. They were their chiefs. Moreover, they could be deified. Their instructions were carried out by people without question. Once, when there was a drought and all Mayan people were threatened with death by starvation, they decided to search for a new place of residence. As we know from the Old Testament, there was a similar case in history. Joseph asked the Egyptian Pharaoh to shelter his people. The Mayans' land was not nearly as prosperous as that of Egypt. From the north, south and east they were surrounded by hostile militant tribes which later conquered them. On the west there was an ocean, and there was a chance that there existed some land in the ocean, or beyond it, which would be suitable for habitation and survival there. They decided to go in search of the "promised" land.

The Mayans had experience of sailing in coastal waters. They loaded their belongings, food and domestic animals onto rafts and, following the guidance of their leaders, set off. Astronomical knowledge helped them to navigate at night by the stars, and at daytime by the sun. About 50 years ago a Norwegian scientist and adventurer, Thor Heyerdahl, with a crew of six people proved that it is possible to cross the ocean from Central America to Polynesia on a raft, "Kon Tiki" [67]. So, such a voyage of Mayans could have been made.

3. Puzzle of Easter Island

> "Imagination is more important than knowledge. Knowledge is limited,
> whereas imagination embraces the entire world, stimulating progress, giving rise to the evolution."
> Albert Einstein

The visit to Chichen Itza made a lasting impression on my wife Anna and me. We often discussed where else on our planet we could find traces of the Mayan tribe, and once we paid our attention to the mysterious Easter Island, towering alone over the boundless waters of the Pacific Ocean. The mystery of the island is in its remoteness from the mainland, almost 2,000 km away, and its intriguing stone statues depicting strange-looking people, spread all over Easter Island. What if the inhabitants of this island were the Mayans' descendants, who fled Central America in prehistoric times? We decided to become acquainted with Easter Island.

Currently, the island is a part of Chile, and to reach it by air, one has to fly from the capital of Chile, Santiago de Chile, which we did. Chile stretches as a long strip from north to south through South America between the high ridges of the Andes and the Pacific expanse, representing almost all climate zones. We met kind, hospitable people with established traditions, rooted in the distant past of their European predecessors. This influenced the country's culture, politics, the political system, etc.

It was the first time Anna and I had visited the Southern Hemisphere. An unfamiliar sky opens up new horizons of the Universe. We observed unfamiliar star patterns, with a genuine interest beheld by the famous constellation of the Southern Cross, which is as significant for navigation in the Southern Hemisphere as the Big and Little Dipper are in the North. A few days later we boarded a plane that took us to Easter Island.

Easter Island is a unique place in the southeastern Pacific Ocean. It is an inhabited island that is one of the most remote from the mainland in the world. It is located 3,703 km west from the nearest coast of the South American continent and 1,819 km east from the nearest inhabited island (Pitcairn Islands). The island has the shape of an isosceles triangle with volcanoes at its vertices, and sides of 16, 18 and 24 km, respectively.

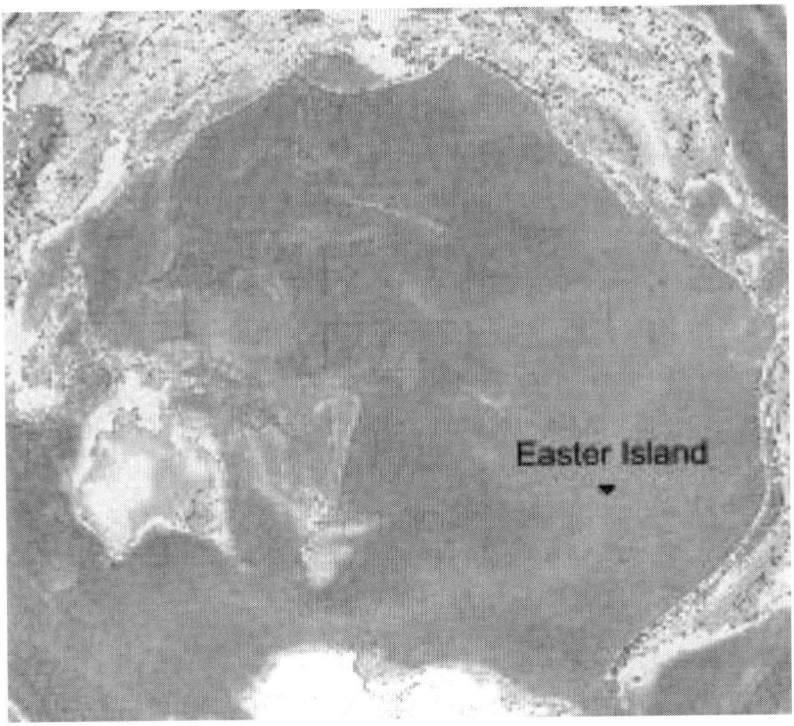

Figure 49. Location of Easter Island in the Pacific Ocean.

The island was discovered by a Dutch explorer, Jacob Roggeveen, on Easter Day, April 5, 1722, and therefore it was named in honor of this holiday. At that time 2,000 locals, who were ethnically similar to South American Indians (or Polynesians similar to them), lived there. According to glottochronological (linguistic) and radiocarbon estimates, the island was inhabited 300-400 years of the first millennium AD, i.e., specifically at the time when the Mayan civilization disappeared.

Figure 50. Central Plaza in Santiago de Chile.

In 1950s, Thor Heyerdahl discovered the recent history of Easter Island. He drew attention to the mysterious island and the statues, suggesting the origin of its inhabitants. In his books, "Kon-Tiki" and "Aku-Aku: The Secret of Easter Island" [68, 69], he assumed that those people had come from Ancient Peru, from the northern part of South America, on the basis of ethnic similarities. He pointed out that the island's statues resemble the statue figurines found in the Andes. Furthermore, yams and sweet potatoes are popular food on the Island and known to have originated from Peru and Colombia. (Note that both of these countries, particularly Colombia, are close geographically, and were possibly in contact with the ancient settlements relating to the Mayan civilization in which we are interested.)

Figure 51. Constellation of the Southern Cross.

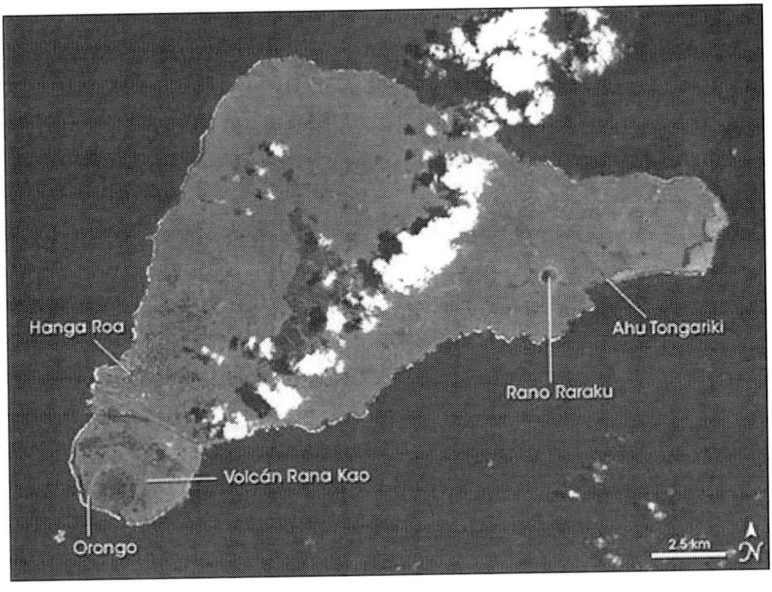

Figure 52. Easter Island.

Figure 53. Anna and I on Easter Island.

On Easter Island we settled in a small but comfortable hotel, situated opposite the "Armada", a Chilean military post, mostly symbolic. The next morning we observed the daily ceremony of raising the national flag, and then went to explore the island. Our tour guide was a young local woman, who had grown up and received an education in the US. When she was 10 years old, her parents had allowed her to go and stay with an American couple in the States who visited Easter Island as tourists. After graduating from university in Los Angeles, she had returned to her parents on the island.

She introduced us to her father, who was making and selling local crafts.

Figure 54. Our guide.

Figure 55. Inhabitant of Easter Island, selling his products. Possibly a Maya descendant?

In our view, the man looked like a typical Mayan, similar to the people we had seen in Chichen Itza. The stories his daughter told us about the island's history and its customs have persistently returned my mind to the intriguing idea of their origin. Judge for yourself. People landed on an uninhabited island. What would they do first? They would build a home, plant something that could be converted into food, organize some primitive farming and so on. And what did the first settlers on Easter Island do? They began to build monolithic human statues (moai) with an incomprehensible obstinacy from rocks (mainly basalt tuff) found on the slopes of extinct volcanoes. The size of the statues ranged on average from 3 to 12 meters, and the weight from 3 to 15 tons. The biggest statue has a size of 20-22 m and a weight of 150-200 tons. It is situated in the quarry and not separated from the base.

Figure 56. Statues on Easter Island.

Now imagine that the Mayans' flotilla arrived safely to the then uninhabited Easter Island. There was some reserved food supply. There were plenty of green fields with enough grass to feed the domestic animals they had brought along. In the volcanic craters there were blue freshwater lakes, and the ocean provided a variety of seafood such as fish, crabs and turtles. In short, survival was possible. But the marshals of the Mayans wanted more. They were looking for "the mainland," the continent. And so far, they had only found a small island lost in the ocean. They decided to continue their search; they took a few people and boarded a raft again, promising to come back for the remaining people once they'd found the mainland. As you will remember, ordinary Mayan people worshipped their marshals. Based on past experience, they had every reason to believe them. The people who remained on the island waited patiently for the return of their voyagers. They started to cut down the slopes of the volcanoes, building statues of their leaders, installing them along the coast and praying for their safe return. They thought that the size of the statutes was essential, so they tried to make them as large as possible, visible from afar. About a thousand such statues can be found today. Some of them are installed along the coastline of the island, a part of them on the slopes of extinct volcanoes; some were exported to Chile, and some of the statutes, not cut down at the base, remained in the land of the quarries.

Whether the leaders returned or not, we do not know. But as an art work in themselves, the statues could have helped their creators to keep their emotional spirits up, i.e., the entropy perturbation, which had guided them in their titanic work on the slopes of three volcanoes on the Easter Island. This disturbance could be felt in close communion with the statues. The statues with the reduced entropy induced an increase of entropy around them (to pick up the standard level of entropy of the statue-people system). As a result, one would feel relaxation, lightness throughout the body, and other symptoms associated with an increase of entropy. Not coincidentally, the locals believed that the statues enclosed the supernatural power of their ancestors.

Figure 57. Statues on the slope of a volcano.

I decided to test it on myself. I came close to the statues and mentally tried to interact entropically with them, imagining that we were forming a mutual thermodynamic system. Please, do not think that I had gone insane! I was carrying out a "pseudo scientific experiment" according to generally accepted notions, trying to justify it "scientifically." I had a feeling of lightness, vitality, well-being (similar to what you feel on visiting a spa). It seemed as if the statue "filled" me with energy. Or maybe it was just my imagination?

Figure 58. My entropic interaction with one of the statues.

Figure 59. The entropic interaction with the head of the statue lying on the ground.

Figure 60. Anna "interacts" with the statue. Compare the sizes! We had experienced similar feelings at the Triangle of Desires in Abadiania and at the Western Wall of Jerusalem in Israel.

Anna and I spent a few days on Easter Island, inspected all of its corners, climbed to the tops of the volcanoes, swam in the ocean, sunbathed on the beach with pristine white sand, and then returned to Santiago de Chile. It has been 8 years since our trip. Little has changed in regards to our knowledge of the Mayan civilization, but an "aftertaste" of this visit, its "flavor", is still palpable.

XII. Noetic Science

(The Greek word "noetikos" means "intuitive mind" or "inner knowing.")

Noetic Science explores the untapped potential of the human brain. From the perspective of "official science," most of the research conducted within Noetic Science refers to "pseudo-scientific experiments." Usually included are the study of the interaction of our consciousness with the environment, and attempts to materialize it, and to use it in the process of medical treatment. Criticism of these studies is most often associated with the reliability of the reported data and the possibility of ambiguous interpretation. What helped to get rid of the disease: the hands of the healer, the medicine or the internal forces of the body itself? Was it possible that the received telepathic information was actually accompanied by some unaccounted types of communications during the experiment? Was it really an act of will by a person or by a group of people, that affected the event, or was it just a pure coincidence?

These and similar questions arise time and again during the comprehension process of the findings of such research. The mission of the Institute of Noetic Science is to prove their authenticity, to give them a reasonable scientific explanation.

The Institute was founded in 1973 by the astronaut of spaceship Apollo 14, Edgar Mitchell. He experienced a kind of shift in consciousness from his flight to the moon and his walks on its surface, so unlike that of the Earth. An acute sense of strangeness regarding what he saw during his flight into space and his moments on the moon's surface marked a turning point in his life. As acknowledged by Mitchell, his mission had become "to expand the knowledge of Nature, the potentials of the mind and consciousness and apply this knowledge to improve human well-being and the quality of life on the planet".

Physicists, biologists, engineers and psychologists work at the Institute of Noetic Science. Their experiments clearly demonstrate the interaction between animate and inanimate Nature, the effectiveness of the methods of alternative medicine, and the endless opportunities of human consciousness. Previously, I described one of the projects carried out by scientists at the Institute ("Global Consciousness Project"). Its results demonstrate the impact of the collective consciousness of people on the world around us. The world reacts to what happens in our minds: our thoughts, emotions, and hopes. The opposite is also true: everything that happens in the world is reflected by a corresponding perturbation of our consciousness.

XIII. Biocentrism and the Thermodynamic Approach to Natural Phenomena

Human impact on the world around us, the "strange" (from the standpoint of classical physics) probabilistic nature of the world, and the fact that the origin of the Universe has not been clarified to till now - all this spurred the prominent American biologist and physician Robert Lanza to create an intriguing theory of the Universe - Biocentrism [70]. Its main thesis states that the human being is not a product of the Universe, but rather the Universe was created by the human mind.

How is it to be understood? A hundred major global constants (such as the weight and dimensions of atoms, elementary particles and their electrical charges, gravitational and electromagnetic constants, strong and weak interactions, the speed of light, and so on) appear to be selected in such a way that makes our world so unique in its design. A change of at least one of these quantities by only one percent, or even less, would unavoidably lead to a catastrophic situation for all biological species of plant and animal origin (and thus for the whole of humankind). Computer simulation of the creation and development of organic life has demonstrated the impossibility of its appearance in even minimally modified conditions. And this is just one of Robert Lanza's arguments.

From the point of view of our thermodynamic approach, the emergence of organic life is a natural one, because the growth of entropy is quicker in the world of animate Nature, than in the inanimate world. Therefore, the tendency to transition into a more probable state is greater in the animate world. As a result, such an evolution of the Universe which the organic World appears is preferable.

"And how about space and time?" Lanza asks "Are they a fruit of the imagination of our consciousness?" The theory of relativity states that time and distance, as measured by us, are not absolute. They depend on the reference system in which we live, and significantly change depending on the speed and nature of its movement, and also on the size of the present gravitational field.

If we are to address the measurements taken in quantum mechanics, the presence of the experimenter and his behavior influence the results, switching on and off the corpuscular and wave mechanism.

All of the objects seen by us are a result of the interpretation by our consciousness of that information which comes to us via photons reflected or radiated by these objects. None of the photons reaching the retina of our eyes gives us information on the form, the size, or the color of the object from which they came. Only the length of an electromagnetic wave or its frequency and energy – that is all that we are told by these photons. But the interpretation of this information by our consciousness allows our imagination to present this object in great detail. "So, what initially exists and is real?" Lanza asks, "The object, or our idea of it?" "The object" materialist philosophers answer. "Consciousness," idealists retort. "The creator," the Bible tells us. In other words, we have no consensus on this important question yet. And it is necessary to take into account the opinion of an undoubtedly talented person, the author of biocentrism.

XIV. Instead of an Epilogue

The thermodynamic approach (outlined in this book) is based on the laws that are common to the Worlds of animate and inanimate Nature. It introduces the concept of the entropic interaction of objects of Nature as the quintessence of Nature. Four fundamental interactions that occur in Nature – gravitational, electromagnetic, strong and weak – are considered as "affiliates" of the entropic interaction, which in conventional current understanding is not entirely new. In a narrow sense, it is used to describe many phenomena in various fields of physics, chemistry, and other sciences; however, it is not regarded as fundamental. It is often mentioned in the subjunctive mood (for example: "the units of the macromolecule appear to entropically repel each other at a close range and are attracted to each other at a longer distance").

In the "thermodynamic approach" the entropic interaction acts as the key action for the transmission of information for many phenomena and processes occurring in Nature. It can be used as a second signal system in the human body, if the central nervous system is damaged and cannot perform its functions in full. It is quite possible that with the help of entropic interactions we can establish contact with alien civilizations.

The entropic interaction is defined as a mutual influence of open thermodynamic systems on the state of each other by changing their summarized entropy and subsequently transferring them to a more probable state. It is regarded as the basis of phase transitions, osmotic effects, dissolving and mixing substances, gravity, quantum entanglement, and of many other phenomena. It explains the interaction of human consciousness with the world surrounding us and with the consciousness of other individuals. With its help, one can relatively easily explain many enigmatic phenomena currently attributed to the category of miracles, or to the category of the far-fetched and mythical. The scientific, spiritual and religious approaches to such phenomena developed to date provide different levels of trust in interpretation and understanding, from being fully credible to complete denial. These phenomena, in particular, include telepathy, clairvoyance, telekinesis, the ability of individuals to influence events and predict them, and the ability to heal people without the use of drugs or the like.

Based on my many years of academic experience, and on my personal meetings with people who have "miraculous powers" and carrying out experiments with them, I do not doubt the existence of these phenomena. Using well-known scientific laws, this book manages to prove the possibility of the existence of these phenomena and offers recommendations for achieving desired results with their help in daily life, sports, business, issues of longevity, the improvement of health and the treatment of various diseases. In particular, recommendations for achieving longevity are based on the evidence of the correlation between the rhythm of time and entropic change.

It was previously mentioned that many statements of the theories of psychoanalysis by Sigmund Freud are based only on one (figuratively speaking) "Children's Syndrome." In the foundation of quantum mechanics there is one fact discovered by Max Planck, that the radiation of light is performed not in a continuous manner, but by certain portions - quanta. The acceptance of the speed of light as a constant in all reference frames enabled Albert Einstein to create the special theory of relativity, and the identification of inertial and gravitational masses, coupled with the interpretation of gravity, as a change in the geometry of space led him to the formulation of the general theory of relativity.

This book shows that most of the phenomena of interest to us may also be substantiated on the basis of only one principle (well known in science), common to the world of animate and inanimate nature. It states that the probability of a given state of an isolated thermodynamic system is completely determined by the amount of energy (called free energy) that the system can spend to do the work. This energy consists of two components: the internal energy (or enthalpy); and the entropy part, which is the product of the entropy of the system, and its temperature in the absolute scale. In the stationary (equilibrium) state, the system has minimal free energy. Any increase of the energy (for example, as a result of external influence) leads the system from a stationary to a less probable state, from which it seeks to go back.

The free energy change can occur due to changes in either of its components. If the system has received additional internal energy from the outside, it can return to a stable state by returning this energy in any way, or by increasing the entropic part of its free energy. Similarly, changing the entropy component of the free energy may be compensated by the corresponding enthalpy changes.

I want to remind you about some of the phenomena discussed in the book. For example, no one doubts the existence of hypnosis, but no one can explain it. But this is a kind of telepathy, i.e., the transmission of a thought at a distance, carried out in visual contact. In a hypnosis session two people form a common thermodynamic system. A disturbance in the brain of one of them, giving a command, lowers the entropy of the entire system and outputs it from the stationary state. To return to this state, the order must be fulfilled, and this happens.

In a standard telepathic experiment, a piece of information is usually transferred without the use of any means of communication. The effect also works as hypnosis. Treatment using the method of contactless therapy also adjoins to these phenomena. Focusing on a patient who is arbitrarily far from the healer, the latter automatically includes the patient into his (her) thermodynamic system and reduces the entropy of the system through a corresponding perturbation of his (her) brain. This happens instantly, i.e., the entropic interactions of the healer and patient are in the same system, tantamount to quantum entanglement. The system leaves its steady state and in order to return to it the organism of the patient should increase its entropy and thereby improve its health. The patient's immune system is activated additionally, the necessary processes are included and the patient is treated more actively.

At contact therapy, when the healer uses his (her) hands, the same mechanism acts as in the non-contact therapy, with the only difference being that the patient experiences an additional lowering of entropy in the places of contact with the healer's hands (and, naturally, throughout the body).

In a telekinetic experiment, the healer is replaced by a psychic (i.e., a person exercising volitional influence), and the patient is replaced by an inanimate object. By a volitional effort the psychic combines himself (herself) with the object into a mutual thermodynamic system. The entropic perturbations arising in the psychic's head displace the entire system from a steady state, and in it there are processes seeking to restore it. Under the influence of these processes, the object begins to move in accordance with the wishes of the psychic. Among the possible processes can be a cluster of ions on the palms of the psychic, the induction of electrical charges on the surface of the selected object and, ultimately, its movement under the influence of the field, which appears as the result of those processes. Everything is simple and there is no mystery!

The influence of a psychic on events is carried out by his (her) entropic interaction within frames of a more complicated system. In this case, in addition to the psychic, the system can include sports teams, military units, and the environment. In accordance with the desire of the psychic – a better implementation of various techniques by athletes, more successful army maneuvers, changes in the desired direction of the atmosphere conditions and the like – can take place.

Clairvoyance is an ability to see objects far beyond our vision, due to the ability to see them through the eyes of a person actually watching them. The clairvoyant, tuning in to this vision, involuntarily forms one thermodynamic system with a real observer of the objects. The clairvoyant's brain is removed from the equilibrium state by the desire to see the selected objects. The same thing happens with the entire system. It tends to return to its original state and returns to it when the desired effect is reached. It occurs by comparison of the pictures imagined by the clairvoyant with real ones, i.e., the loss of the entropic perturbation in the observer's brain.

At the coincidence of the pictures, the clairvoyant experiences an abrupt increase of his entropy to the equilibrium level and his (her) imagination conjures up the proper picture.

Mirages can be explained in the same way, only in this case the apparent pattern can also be initiated by the human brain itself. Doesn't this explanation look very "biocentric"?An experiment was conducted in the Noetic Institute in which a researcher (engineer-physicist) located in San Francisco was trying to see what was happening in a Soviet military base in Siberia, located 10-15 thousand miles away. He was guided by the idea that if he could (by his force of will) concentrate on one bunch of the photons flying from this object of interest, he would be able to see the object properly. The experimental result was negative, but the fact of this experiment itself can be seen as both a recognition of the possible existence of clairvoyance, and as an attempt to explain it as an alternative to the thermodynamic entropic approach.

The thermodynamic entropic approach allows for the substantiation of a brief appearance of negative gravity in the initial period of the Universe formation at the "Big Bang." The approach explains the fact that particles of matter were formed so close together that their entropic repulsion exceeded their gravitational attraction.

The entropic approach also predicts an additional possibility for the formation of Black Holes.

The possibility of the existence of a large number of Universes (or, at least, of galaxies in our Universe) resolves the issue of there being among them such Universes that implement a set of universal constants that are favorable for our existence, and it covers the problem of their origin, posed by biocentrism.

The thermodynamic approach quite simply explains the origin and development of life on our planet. On the one hand, the formation of complex molecular structures led to a decrease in entropy. On the other hand, the repetition of these structures and the associated increase in their final size increased the entropy, which is a countably additive function. In addition, the union of simple molecules into more complex ones was accompanied by a decrease of the internal energy (since the binding energy is always negative). As a result, the free energy of the resulting macromolecular compounds decreased, making this development from simple to complex more likely than the separate existence of simple molecules.

The "thermodynamic approach" allows a hypothetical opportunity of establishing a contact and an information exchange with extraterrestrial civilizations by telepathy and clairvoyance in real time without any delay.

There are not too many people with a psychic or healing natural gift. Not everyone is able to influence events; not everyone can treat people using the above-described methods of alternative medicine. It is possible to assume that the ability to perform this is somewhat put in each of us, but that not everyone can use it.

It is known that only a small part (5-15%) of the gray matter of our brain is active. The rest, the bigger part, is idle. This part of the gray matter is seemingly not switched on and is a kind of ballast in our cranium. It is hard to imagine that Nature has created it for future use. Nature is usually economical and creates nothing superfluous. (An impressive example of this is the principle of least action.) Perhaps, at the dawn of mankind, this part of the brain was also actively working, and then, under the influence of external circumstances (such as radiation) it has mutated and turned off. It is natural to assume that the amount of working gray matter varies from person to person. Some people have more of it; some less. It is possible that the ratio of working and non-working gray matter can be inherited, or may develop in the process of life. Neurologists have found that in early childhood (up to five years of age) a significant portion of the brain's neurons are turned on to work. Then, with age, their turning on decreases exponentially. The greater the number of neurons involved, the more information a person can learn, the more advanced their memory and the greater their intellectual abilities.

Turning on neurons can be managed! The more information a child receives at an early age, the more his (her) neurons are turned on!

Parents, remember this! "Invest" in your children as much information as possible – and the sooner the better!

It is also possible that some parts of the brain do not work for most people, but work for the chosen individuals. They may be responsible for functions related to excessive sensitivity to external influences, to an enhanced ability to generate electromagnetic radiation in certain wavelength ranges, to use the entropic signaling system, and hence to react to and affect the thermodynamic state (in particular, the degree of order) of various systems, including both biological and non-biological, and thus to cure humans and affect events.

The World in which we live is lovely! It is full of "miracles"! But I put that word in quotes, because we can find a reasonable scientific explanation for each "miracle":

If you have received appropriate genetic inheritance, possess a strong will and unwavering faith in your abilities, these qualities can be developed, and then the World will be at your feet!

APPENDICES

APPENDIX 1. THE PRINCIPLE OF LEAST ACTION

In the principle of least action, the action S is understood as a functional [4], i.e., a number to associate to each function q(t):

$$S(q) = \int_{t1}^{t2} L(q(t), \frac{\partial q}{\partial t}, t)dt \qquad (1)$$

where $L(q, q_t, t)$ is the Lagrange function of the object, which characterizes its energy and is dependent on time t, on generalized coordinates q (independent variables needed to fully describe the state of the system) and their first derivatives according to time $q_t = \partial q / \partial t$. The principle of least action requires that the evolution of the object leaves unchanged the action functional S, i.e.,

$$\delta S = 0 \qquad (2)$$

Applying the principle of least action to the objects of classical mechanics, one should choose for them a Lagrange function that depends only on the mechanical characteristics of the objects. Suppose, for example, the object in question is in a potential field, where it is subjected to a force F. Then, if the

Lagrangian function for this object is selected as the sum of its potential and kinetic energy,

$$L = Fq + mq_t^2/2 \tag{3}$$

then the Second Law of Newton follows at once from the requirement of equality to zero of the variation of action S(q) that is defined by equation (1)

$$F - mq_{tt} = 0 \tag{4}$$

or

$$F = mq_{tt} \tag{5}$$

The transition from mechanics to thermodynamics can be made if the Lagrangian variation is considered to depend only on the thermodynamic characteristics [2, 6, 7]. Then the principle of least action is reduced to the requirement that the variation of the action vanish at variations in entropy, enthalpy and temperature.

Using the Lagrangian as the sum of potential, kinetic and free energy,

$$L = Fq + mq_t^2/2 + G \tag{6}$$

it is easy to get an equation well known in thermodynamics

$$\delta G = \delta H - T\delta s - s\delta T = 0 \tag{7}$$

reflecting the fact of the mutual compensation of changes in entropy and enthalpy.

Appendix 2. Probability of Having a Macromolecule in the Space between Two Impermeable Membranes

The probability of the macromolecule being in an ideal solvent in the space between two impermeable membranes is easily calculated if every macromolecular conformation is considered as a path of random walk of the monomeric units of the macromolecule. Under appropriate initial and boundary conditions, this probability, W, is expressed by the following function [36].

$$W = 8/\pi^2 \sum_{m=1}^{\infty} 1/(2m+1)^2 exp[-((2m+1)^2\pi^2)/4)((\overline{R^2})^{1/2}/r)^2] \quad (1)$$

where r is a distance between membranes, and $(\overline{R^2})^{1/2} = (l^2 n/6)^{1/2}$ is a mean square radius of inertia of the macromolecule consisting of n of links by length of l.

Obviously, the narrower the space between membranes, the less the probability of a macromolecule being there. The probability of the detection of units of a macromolecule near walls is always less than at a distance from them. Thus, it is possible to say that the macromolecule, seemingly entropically, is being repelled from the walls of the impermeable membrane.

Appendix 3. Proof of the Correlation between Change of Entropy and the Time Rhythm

Consider an isolated thermodynamic system in which there is a certain process. At the time t_1 the state of the system is characterized by a set of physical characteristics. Let the probability of this state be W. As a result of what is happening in a system, changes in its state occur. The probability of this change ΔW at a short time interval is proportional to the probability of a given state W, and time, which is

$$\Delta W = -\nu W \Delta t \quad (1)$$

A minus sign here means that after a time Δt the probability of remaining in the initial state of the system (i.e., with the same characteristics) will decrease.

Based on the equation (1), we can write the differential equation (2):

$$\Delta W / W = -\nu \Delta t, \quad (2)$$

or

$$dW / W = -\nu dt, \tag{3}$$

which has to be solved with the following initial and final conditions

$$W(t_1) = W_1, W(t_2) = W_2 \tag{4}$$

and after integration, we have

$$\int_{t1}^{t2} dW / W = -\nu \int_{t1}^{t2} dt \tag{5}$$

Then

$$lnW_1 - lnW_2 = \nu(t_2 - t_1) \tag{6}$$

where v is a coefficient of proportionality.

As for an isolated system, the logarithm of the probability of its state, taken with the opposite sign, is (according to the H-theorem of Boltzmann) the entropy s of the state (up to a factor of 1/k); we can rewrite this equation as

$$s_2 - s_1 = \Delta s_{2,1} = k\nu(t_2 - t_1) = k\nu\Delta t \tag{7}$$

and

$$\Delta t = (1 / k\nu)\Delta s_{2,1}, \tag{8}$$

where k is the coefficient of proportionality, equal to the Boltzmann constant.

So we arrive at the following conclusion:

Any process that takes place in an isolated thermodynamic system during time Δt change the entropy of the system by an amount $\Delta s_{2,1}$ proportional to the time interval Δt. i.e., changes in the entropy and time are related linearly and proportionally. Moreover, the entropy change is positive, since the value of $\Delta t = t_2 - t_1$ is always positive (i.e., the entropy in equation (8) is increasing with time, as needed, based on the second law of thermodynamics).

However, there is also another opportunity for a change of entropy. If, at deriving the equation (1), we consider that the probability W for the system to stay in an initial state will increase after the expiration of time Δt, or in other words, if the minus sign changes to positive in the right part of this equation, then in solving it, we will receive

$$s_1 - s_2 = \Delta s_{1,2} = k\nu\Delta t > 0 \quad (9)$$

I.e., at such evolution of thermodynamic system its entropy decreases over time (s2 < s1), but time thus continues to increase (Δt > 0).

APPENDIX 4. CORRELATION BETWEEN CHANGE OF ENTROPY AND THE TIME RHYTHM FROM THE PRINCIPLE OF LEAST ACTION

Let us consider from the viewpoint of the principle of least action the behavior of the system of mass m, moving with velocity q_t = dq/dt in the force field F. In this case, the action S is understood as a functional; that is, a number that associates with each function q(t):

$$S(q) = \int_{t1}^{t2} L(q(t), \partial q/\partial t, t)dt \quad (1)$$

where the Lagrange function of the object, L(q(t), ∂q/∂t, t), depends on the time t, the generalized coordinates q and their first time derivatives of q_t = dq/dt. If the object has n degrees of freedom, the Lagrangian depends on the time t, n generalized coordinates $q_i(t)$, i= 1,2, ... n, and their time derivatives.

The principle of least action requires that the first-order variation of the action is zero for all possible disturbances, i.e., the evolution of the object must be such that the action functional, S, remains unchanged.

$$\delta S = 0 \quad (2)$$

This requirement may be satisfied only if it is done by describing the evolution of the object using the Euler-Lagrange equation

$$\partial L/\partial q - d/\mathbf{dt}(\partial L/\partial q_t) = 0, \quad (3)$$

which is obtained by variation of the action functional (1) with the initial and boundary conditions (4):

$$\boldsymbol{\delta} q(t_1) = \delta q(t_2) = 0 \quad (4)$$

Let us choose as Lagrange function, L, the sum of its potential energy, Fx, that is, the energy of interaction with an external field, and the Gibbs free energy, G, entropy part of which is sT, and enthalpy includes the energy associated with the mass of the rest of the system m_0 and its kinetic energy is $m_0c^2/\sqrt{(1-x_t^2/c^2)}$, and the amount of heat in it is Q:

$$L = Fx + G = Fx + m_0c^2/\sqrt{(1 - x_t^2/c^2)} + Q - sT \quad (5)$$

Let the variable characteristics of the system be the spatial coordinate x, speed x_t, entropy s, and heat Q. The requirement (2) for the stationary action S leads to the equation

$$\delta S = \int_{t1}^{t2} [\delta x \partial L/\partial x + \delta x_t \partial L/\partial x_t + \delta s \partial L/\partial s + \delta Q \partial L/\partial Q] dt = 0 \quad (6)$$

Applying integration by parts to the second term in the integral (6), we obtain (with the initial and final conditions (4)):

$$\boldsymbol{\delta} S = \int_{t1}^{t2} [\delta x \frac{\partial L}{\partial x} - \delta x \frac{d}{dt}(\frac{\partial L}{\partial x_t}) + \delta s \frac{\partial L}{\partial s} + \delta Q \frac{\partial L}{\partial Q}] dt = 0 \quad (7)$$

Substituting in equation (7) the expression for the Lagrangian (5), we obtain

$$F\delta x - \delta x[m_0 x_{tt}/\sqrt{(1-x_t^2/c^2)^3} + m_0 x_{tt}(3x_t^2/c^2)/(1-x_t^2/c^2)^{5/2}] + \delta Q - T\delta s = 0$$

This means

$$\delta s = \left\{ F\delta x - \delta x[m_0 x_{tt} / \sqrt{(1 - x_t^2 / c^2)^3} + m_0 x_{tt}(3x_t^2 / c^2)/(1 - x_t^2 / c^2)^{5/2}] + \delta Q \right\} / T \quad (8)$$

.

That is, the choice of the Lagrangian according to equation (5), in a coordinate system moving with the velocity x_t in the force field F, the entropy of a thermodynamic system is dependent on velocity x_t, on the rate of acceleration x_{tt},, the value of F, and changes in the amount of heat in the system. Here, m is the mass of a thermodynamic system, moving at speed x_t and having a value of the state of rest: $m = m_0 / \sqrt{(1 - x_t^2 / c^2)}$, c is the speed of light. The existence of the speed of movement x_t and an acceleration x_{tt} reduces the entropy and influence of a force field (term $F\delta x$ in (8)) increases it if this field is a potential one and the force of F acts as pushing away from a field source. If this force works as an attraction, i.e., a field, for example, is a gravitational field, the value of in the equation $F\delta x$ in the Eq.8 is a negative value and leads to the reduction of entropy. The entropy of the system also grows more slowly with an increase in the mass of the system, which is reflected by the multiplier in the second term of the equations (8). The removal of some amount of heat from the system ($\delta Q < 0$) leads to the reduction of entropy as well ($\delta s < 0$).

All of the above-mentioned effects are inversely proportional to the system temperature T: with the temperature reduction they are intensified, i.e., the reduction of entropy becomes more essential.

However, it would be incorrect to think that the entropy and time changes identically coincide. According to the principle of causality, the processes occurring in Nature can be divided in time so that the previous ones can have an impact on the subsequent ones. This cannot be said about the entropic changes connected with these processes, i.e., it is impossible to claim that for the isolated thermodynamic system the state with smaller entropy preceded the state with higher entropy. Unlike the time, entropic changes have a statistical nature. Each closed thermodynamic system (if it is brought out of an equilibrium state) seeks to pass to a state of greater equilibrium (i.e., a more probable state), though it can also pass to a less probable state (i.e., less equilibrium). The entropy upon such transition will decrease, and the time will continue to increase, without changing its direction. This opportunity is very small, but nevertheless it exists and can be estimated according to Poincaré's recurrence theorem [20].

It is possible to come to this opportunity also on the basis of the equation (1) from Appendix 3 if we consider that the probability W for the system to stay in the initial condition will increase after time Δt. In this case, it is necessary to change the minus sign to plus in the right side of this equation. As a result, in solving it, we obtain:

$$s_1 - s_2 = \Delta s_{1,2} = k\nu\Delta t > 0 \tag{9}$$

i.e., at such evolution of the thermodynamic system, its entropy decreases over time

($s2 < s1$), but the time continues to increase ($\Delta t > 0$).

APPENDIX 5. ENERGETICS OF THE PROCESS OF THE FORMATION OF A MOLECULE CONSISTING OF TWO ATOMS

Let us consider the simplified process of the formation of the molecules of two atoms of different chemical elements. Let $H_{1,1}$ be the energy of a contact between two atoms from the solvent, which contains minerals with these elements. And let $H_{2,2}$ and $H_{3,3}$ be the contact energies of the atoms in each of the minerals. For the dissolution of one atom of each of the minerals (with the energies of their dissolution, $H_{1,2}$ and $H_{1,3}$), their contacts with other atoms of minerals have to be severed. There should also be severed contacts of solvent atoms with each other (with the energy $H_{1,1}$). In the end, a contact with the energy $H_{2,3}$ may form between these freed atoms of the minerals. The energy of this contact will consist of the dissolution energies of these atoms

$$\Delta H_{1.2}=0.5(H_{1,1}+H_{2,2})-H_{1,2} \tag{1}$$

$$\Delta H_{1.3}=0.5(H_{1,1}+H_{3,3})-H_{1,3} \tag{2}$$

and their subsequent interaction leading to the formation of new molecules.

$$\Delta H_{2.3}=0.5(H_{1,2}+H_{1,3})-(H_{2,3}+H_{11}) \tag{3}$$

The internal energy of a thermodynamic system H decreases during the interaction, which leads to the structuring and the corresponding decrease in entropy. In our example, the union of two atoms in one molecule is always associated with a loss of internal energy ΔH. If the loss is greater than the loss of the entropic term of Gibbs free energy, $T\Delta s$, then the formation of such a molecule is preferable to the separate existence of the considered atoms. The stronger the interaction binding the atoms in the molecule, the greater the probability of its formation.

References

[1] Einstein, Albert, 1905, On a Heuristic Point of View Concerning the Production and Transformation of Light, *Annalen der Physik* 17(6): 132–148.

[2] Einstein, Albert, 1905, On the electrodynamics of moving bodies *Annalen der Physik* 17(10), 891-921.

[3] Einstein, Albert, 1916, The Foundation of the General Theory of Relativity*, Annalen der Physik* 49 (4), 769-822, 1916.

[4] Guth, Alan H. 1997, The Inflationary Universe, Reading, MA: *Perseus Books.*

[5] Hawking, Stephen W. & Ellis, George Francis Rayner, 1973. The Large Scale Structure of Space–Time. Cambridge University Press.

[6] Freud, Sigmund, 1989, Introductory Lectures on Psychoanalysis, WW Norton & Co. Inc.

[7] Landau L. D. and E. M. Lifshitz, 1976, Theoretical physics, Vol, 5, Statistical physics, part 1, *Nauka,* Moscow.

[8] Hume, David 1896, A Treatise of Human Nature, Clarendon Press.

[9] Born, M. 1949, Natural Philosophy of Cause and Chance, Oxford.

[10] Bohm, David. 2005. Causality and Chance in Modern Physics. London: Taylor and Francis.

[11] Gilbert D., 1923, Basis of Geometry, Translation from German, L.

[12] Mach, Ernst, 1960, The Science of Mechanics; a Critical and Historical Account of its Development. LaSalle, IL: Open Court Pub. Co.. LCCN 60010179. This is a reprint of the English translation by Thomas H. MCCormack (first published in 1906) with a new introduction by Karl Menger.

[13] PLM de Maupertuis, 1744, *Mém. As. Sc.* Paris p. 417.

[14] Einstein, Albert, letter to Ernst Mach, Zurich, 25 June 1923, in Misner, Charles; Thorne, Kip S.; and Wheeler, John Archibald (1973). *Gravitation.*

[15] Arfken,G., 1985, Mathematical Methods for Physicists, 3rd ed. Orlando, FL: Academic Press.

[16] Sommerfeld A., 1955, Thermodynamics and statistical physics, translation from German, M.

[17] Kondepudi, D. & Prigogine, I. 1998, Modern thermodynamics. New York, NY: *Wiley.*

[18] Brown, M. 2005. editor, Feynman's Thesis: The Principle of Least Action in Quantum Mechanics, *World Scientific,* Singapore.

[19] Kibble, Classical Mechanics, European Physics Series, McGraw-Hill (UK), 973.

[20] Maslov, V. N., 1991, Thermodynamic basis of the principle of least action. Izvestiya Vysshikh Uchebnykh Zavedenii, *Fizika,* (Russ. Phys. J.) 5, 53–59.

[21] Hanc, J. & Taylor, E. F., 2004, From conservation of energy to the principle of least action: a story line. *Am. J. Phys.* 72, 514–521.

[22] Ritz, W. & Einstein, A. 1909, Stand der Strahlung Problems. *Zum gegenwartige. Phys. Z.* 10, 323–324.

[23] Hilbert, D.,1915, Die Grundlagen der Physik Konigl. Gesell. d. Wiss. Göttingen, *Nachr. Math.-Phys. Kl.* 395-407.

[24] Feynman Richard P., 1995, Feynman Lectures on Gravitation, Addison-Wesley, p. 136, eq. (10.1.2); Feynman, RP & Hibbs, AR 1965, Quantum mechanics and path integrals. New York, NY: McGraw-Hill.

[25] Carroll, Sean M. 2004, *Spacetime and Geometry,* Addison Wesley.

[26] Huggins M., 1942*, J. Phys. Chem.,*46, 161.

[27] Flory P. J., 1942*, J. Phys. Chem.,*10, 51.

[28] Bresler S. E. and Erusalimsky B. L., 1965, Physics and chemistry of macromolecules, *Nauka,* M.-L.

[29] Wolkenstein M. W., 1959, *Configuration statistics of polymeric chains*, Prod. Academy of Sciences of the USSR.

[30] Israelachvili J., 1985, Intermolecular and Surface Forces. London: Academic Press.

[31] Schrödinger E., 1935, Discussion of Probability Relations between Separated Systems, Proceedings of the Cambridge Philosophical Society: № 31. p. 555.

[32] Einstein A, Podolsky, B., Rosen N. 1935, *Physical Review* 47, p. 1804.

[33] Peres Asher, 1993, Quantum Theory, Concepts and Methods, *Kluwer,* p. 115.24.

[34] Laing A, Lobino M, Matthews JCF, Thompson MG, O'Brien JL, 2012 "Generating, manipulating and measuring entanglement and mixture with a reconfigurable photonic circuit." *Nature Photonics,* 6: 45–59.

[35] Kroto H. W., Heath J. R., O'Brien S. C., et. al. 1985, C_{60}: Buckminsterfullerene, *Nature* 318, 162.

[36] Casassa E, 1969, *Macromolecules,* 2,14.

[37] Belenky B. G. and Vilenchik L. Z., 1983, Modern liquid chromatography of macromolecules, *Elsevier.*

[38] Holler, F. James; Skoog, Douglas A; Crouch, Stanley R., 2007. "Chapter 1." *Principles of Instrumental Analysis* (6th ed.). Cengage Learning. p. 9.

[39] Birkhoff, George David, 1942, "What is the ergodic theorem?," American Mathematical Monthly (The American Mathematical Monthly, Vol. 49, No. 4) 49 (4): 222–226.

[40] Newton I., 1915, Mathematical principles of the natural philosophy, Ps., p. 30.

[41] Einstein, Albert, 1966 Time, space and inclination, Collection of scientific works. Volume 2, p. 715, Moscow, *"Science."*

[42] Barreira, Luis, 2006, "Poincaré recurrence: old and new." In Zambrini, Jean-Claude. XIVth International Congress on Mathematical Physics. *World Scientific*. pp. 415–422.

[43] Vilenkin Alex, 2007, Many Worlds in One, Hill and Wang. New York.

[44] Oort, J. H., 1932, "The Force Exerted by the Stellar System in the Direction Perpendicular to the Galactic Plane and Some Related Problems," *Bull. Astron. Inst. Neth.* 6, 249-87.

[45] Freeman Ken, Geoff McNamara, 2006. In Search of Dark Matter. *Springer.* ISBN 978-0-387-27616-8.

[46] Zwicky, F. (1933), Die Rotverschiebung von extragalaktischen Nebeln, *Helvetica Physica Acta* 6: 110–127.

[47] Rubin, Vera C.; Ford, W. Kent, Jr., 1970, "Rotation of the Andromeda Nebula from a Spectroscopic Survey of Emission Regions." *The Astrophysical Journal* 159: 379. Bibcode:1970 ApJ...159..379R. doi:10.1086/150317.

[48] Einstein, Albert and Infeld, Leopold, 1938, Evolution of Physics. Cambridge University Press.

[49] Hawking, Stephen 1988. *A Brief History of Time,* Bantam Books, Inc.

[50] Schrodinger Ervin, 1992, *What is Life,* Cambridge University Press.
[51] England, J. 2013, "Statistical Physics of self-replication." *J. Chem. Phys.*, 139, 121923.
[52] Bruice Paula Y., 2007, Organic Chemistry (5th Edition), Saddle River, NJ: Pearson Prentice.
[53] Yingjie Li Yingjie et al., 2010. "Fabrication and evaluation of chiral monolithic column modified by β-cyclodextrin derivatives." *Talanta* 80 (3): 1378–1384.
[54] Iorish Yu., Turobov B., 1984, *"Priroda."* No. 11.
[55] Shpunt V. Kh. et al. 1995, Electronic Phenomena in Chalcogenide Glassy Semiconductors, *Nauka Publishers*, St. Petersburg, Russia, 500 pages.
[56] Shpunt V. Kh., Rud Yu. V. and Tsendin K. D., 1993, Optical and Electrical Properties of Human Skin, *Tech. Phys. Lett.,* Vol 19, p. 415.
[57] Shpunt V. Kh., Shifrin E.I., and Tsendin K. D., 1991, Long-term Integral Electrical Memory of the Human Body, *Sov. Tech. Phys. Lett.,* Vol 17, p.93.
[58] Nelson Roger and Bancel Peter, 2011, Effects of Mass Consciousness: Changes in Random Data During Global Events, EXPLORE, November/December, Vol. 7, No. 6, 373-383.
[59] Radin Dean, 1977, The Conscious Universe, *Harper One.*
[60] Radin Dean, 2006, Entangled Minds, Paraview Pocket Books.
[61] Guy David and Shihman Igor, 2008, Working wonders, Shihman Publishing, New York.
[62] Kosmann-Schwarzbach, Yvette, 2010, The Noether theorems: Invariance and conservation laws in the twentieth century. Sources and Studies in the History of Mathematics and Physical Sciences. *Springer-Verlag.*
[63] Oerter R., 2006, The Theory of Almost Everything: The Standard Model, the Unsung Triumph of Modern Physics (Kindle ed.). *Penguin Group.*
[64] Kaku M., 2008. Physics of the Impossible: A Scientific Exploration Into the World of Phasers, Force Fields, Teleportation, and Time Travel., *Doubleday Publishing.*
[65] Bais S., 2005, The Equations: Icons of knowledge. *Springer,* p. 84-88.
[66] Levine, Ira. N 1978. Physical Chemistry, University of Brooklyn: *McGraw-Hill.*
[67] Landsberg. P. T, 1978. Thermodynamics and Statistical Mechanics, *Oxford University Press,* Oxford UK.
[68] Heyerdahl, Thor,1950. Kon-Tiki. Rand McNally & Company.

[69] Heyerdahl, Thor, 1958, Aku-Aku: The Secret of Easter Island, Rand McNally & Co.
[70] Lanza Robert and Berman Bob, 2009, *Biocentrism, Benbella Books*, Inc., Dallas.

INDEX

A

B

C

D

E

F

G

H

I

J

K

L

M

N

O

P

Q

R

S

T

U

V

W

Y